Mounira Ben Slimane
Rym Bouhlal
Hager Snoussi

Wine Heritage and Traditions in Kerkennah

Mounira Ben Slimane
Rym Bouhlal
Hager Snoussi

Wine Heritage and Traditions in Kerkennah

ScienciaScripts

Imprint

Any brand names and product names mentioned in this book are subject to trademark, brand or patent protection and are trademarks or registered trademarks of their respective holders. The use of brand names, product names, common names, trade names, product descriptions etc. even without a particular marking in this work is in no way to be construed to mean that such names may be regarded as unrestricted in respect of trademark and brand protection legislation and could thus be used by anyone.

Cover image: www.ingimage.com

This book is a translation from the original published under ISBN 978-620-6-70485-0.

Publisher:
Sciencia Scripts
is a trademark of
Dodo Books Indian Ocean Ltd. and OmniScriptum S.R.L publishing group

120 High Road, East Finchley, London, N2 9ED, United Kingdom
Str. Armeneasca 28/1, office 1, Chisinau MD-2012, Republic of Moldova, Europe
Printed at: see last page
ISBN: 978-620-7-71220-5

Copyright © Mounira Ben Slimane, Rym Bouhlal, Hager Snoussi
Copyright © 2024 Dodo Books Indian Ocean Ltd. and OmniScriptum S.R.L publishing group

Happy

Mounira Ben Slimane, Rym Bouhlal, Hager Snoussi
Emna Jedidi, Sonia Bouhachem, Rebha Souissi
Mehdi Trad & Naïma Mahfoudhi

Institut
National de
La Recherche
Agronomique
de Tunisie

Université
de Carthage
Tunisie

2024

Heritage and Wine Traditions in Kerkennah

Mounira Ben Slimane, Professor of Higher Agricultural Education, Viticultural Genetic Resources, Horticulture Laboratory, National Institute of Agricultural Research of Tunisia. E-mail : mounira.bsh@gmail.com

Rym Bouhlal, Assistant of Higher Agricultural Education, Ecophysiology and pomology of fruit trees, Horticulture Laboratory, National Institute of Agricultural Research of Tunisia.
E-mail : rym.bouhlal@gmail.com; rym.bouhlal@inrat.ucar.tn

Hager Snoussi, Lecturer in Higher Agricultural Education, Biotechnology, Molecular Genetics of Plants, Horticulture Laboratory, National Institute of Agricultural Research of Tunisia.
E-mail : hagersnoussi@gmail.com ; snoussi.hajer@mratucar.tn

Emna Jedidi, Assistant Professor of Higher Agricultural Education, Vegetable Biotechnology, Tissue Culture, Horticulture Laboratory, National Institute of Agricultural Research of Tunisia.
E-mail: jedidi.emy@gmail.com

Sonia Bouhachem , Professor of Higher Agricultural Education, Entomology, Plant Protection Laboratory, National Institute of Agricultural Research of Tunisia.
E-mail : bouhachems@gmail.com; bouhachem.sonia@inratucar.tn

Rebha Souissi, Senior Technician, Entomology, Plant Protection Laboratory, National Institute of Agronomic Research of Tunisia.
E-mail: srebha@yahoo.com

Mehdi Trad , Assistant Professor of Higher Agricultural Education, Quality and Post-harvest Physiology of Fruits, Regional Center for Research in Horticulture and Organic Agriculture. E-mail : mh.trad@yahoo.com

Naima Mahfoudhi, Professor of Higher Agricultural Education, Plant Virology, Plant Protection Laboratory, National Institute of Agricultural Research of Tunisia. E-mail: naimmahfoudhi@gmail.com

Thanks

The work "Heritage and Wine Traditions in Kerkennah", presented here, partly testifies to the incredible wine-growing wealth of the Kerkennah Islands. This book presents valuable information on island vines in their growing sites. This rich diversity, making Kerkennah residents and lovers of the island proud, faces numerous challenges including climate change and anthropogenic activities which raise fears of an irreversible loss of Kerkennah vines, agricultural and wine-growing traditions and particularly that of the Asli grape, the island's flagship variety.

This work is the fruit of the efforts of a dynamic and diverse team of specialists and Kerkennah enthusiasts whom we would like to thank.

Thanks to Mr. Habib Ben Cheikha, Director of the Territorial Extension Unit (CTV) of Kerkennah who accompanied us during all the surveys and monitoring of the cultivation of vines on the ground.

We would like to thank Messrs. Nejib Megdiche and Nejib Kachouri for their warm welcome and their availability to assist us during our field visits.

A special thought to Mrs. Rachida Ben Ezzeddine for her warm welcome. She taught us a lot about traditional culinary techniques using the Asli grape.

We would also like to thank Mr. Marwen Azzabou for passing on his know-how to us regarding the preparation of "Assir" using the traditional Kerkennian technique.

We thank all the Kerkennian farmers with whom we had a great pleasure working together. They received us well and taught us a lot about the traditional techniques of vine growing, its transformation and its uses.

Miss Cosima Bassouls, during the preparation of her end of studies diploma at the Ecole Supérieure d'Agro-Development International (ISTOM, France), under the direction of Prof. Mounira Ben Slimane (INRAT), contributed significantly effective in carrying out surveys in the Kerkennah Islands, on the distribution of grape varieties and their associations with particular attention to the distribution of ancestral cultivation areas of the star variety "Asli" as well as its cultural practices.

We would like to thank Mr. H'mida Ben Hamda, principal engineer at the National Institute of Agronomic Research of Tunisia (INRAT) for his valuable help during field reconnaissance missions, as well as for sharing with us his expertise and knowledge of island agricultural systems.

Preface

The Kerkennah Islands, a splendid place, have already attracted the attention of many people passionate about nature. The "Vineyards of Kerkennah", the "Grapes of Kerkennah", the "Assir" ... are terms often repeated for a long time and have sparked a lot of discussion. More or less old writings, relatively recent publications, conferences and training and information days have dealt with these subjects. So, what can this new work add? A concerned scientist wanted to represent the theme in her own way.

Its presentation is better anchored in the history of the Kerkennian vineyards. These are observed in space and time by the eye of a passionate researcher. Ms. Mounira Ben Slimane and her collaborators explore, discover and adequately illustrate a very rich and diversified heritage from a perspective of rehabilitation and improvement. In describing this heritage, the authors consider the physical, biological and socio-economic environment. Beyond the diversity of grape varieties (varieties), the Kerkennah vineyard attests to a diversity of cultural practices, culinary habits and very interesting social attitudes. All these components have found their place in this writing.

To what has already been written by predecessors, the authors correctly describe the main grape varieties (Asli and many others). They want to go beyond a catalog, a classic ampelography. to relive the good times of Kerkennah. This work helps support local winegrowers from the choice of grape variety to the harvests celebrated until today in certain villages. He takes us to the "wine growers" who tend their orchards and brings us to the local "enologists" in their "artisanal cellars".

The notion of biological resources or biodiversity is no longer the one we had used until now. It was necessary to open up other socio-cultural and socio-economic dimensions.

This work invites us to appreciate and contemplate this biological, technological and sociological heritage. Understanding this history is, in fact, essential to imagine and prepare for the future of this sector, the driving force behind sustainable agricultural development of the islands.

Messaoud MARCH
Professor, Production Sciences
Plants and the Environment

Summary

The work "Heritage and Wine Traditions in Kerkennah", is an exhaustive study which highlights the history of the vine and the wine culture of the Tunisian archipelago of Kerkennah, a particular and difficult environment, witness to the history of civilizations of the Mediterranean, which despite its vulnerability remains highly coveted.

The authors describe the vines in the historical and environmental context of the Kerkennah terroir. The vine, this valuable genetic heritage forming an integral part, not only of the agricultural production system but also of the rich history of the archipelago. Despite pedoclimatic constraints such as drought, degradation of plant cover, desertification, soil salinization, and climate change, the vine introduced by the Carthaginians persists, tracing the patterns of evolution of plots and management methods. practice.

The work then explores the traditional methods and know-how as well as the cultural practices of the vine and the production of fresh grapes, dried grapes and wine, with a special mention to the flagship indigenous grape variety of Asli Island, used for fresh consumption, drying as well as local artisanal winemaking. The multiplication and physiological characterization of the development of leaves and buds of the Asli grape variety are described and a detailed study of its fresh and dried grapes and its wine as well as its use in Kerkenian cuisine is presented. Both endowed with exceptional quality of fruits and their derivatives but also with a capacity for adaptation and resilience to increasingly severe aridity. The different varieties of Kerkennah vines constitute a reservoir of genes of undeniable interest. They are described in this work and their variability evaluated by ampelographic, biochemical, cytological and molecular analyses. Varietal sheets are created for the eight representative varieties of the wine varietal assortment: Asli, Mehdoui, Jerbi, Kohli, Marsaoui, Hamri, Dalia and Tounsi. In addition to these sheets, there are plates illustrating the adult leaf, the budding, the cluster, the flower and the seeds, the essential organs of the variety.

The ampelographic description is carried out according to the codification of the International Union for the Protection of New Varieties of Plants (UPOV), the International Council on Genetic Resources (IPGRI) and the International Organization of Vine and Wine (OIV), describing the main quantitative and qualitative distinctive characteristics of the different organs of the vine plant considered at specific stages of vegetative development. The flow cytometry technique allowed cytological characterization; the chromosomal enumeration of representative varieties of indigenous cultivated vines of Kerkennah attests to their diploidy (2n=38).

Molecular analysis by nuclear microsatellites revealed a strong genetic diversity of the vine varieties analyzed as well as an allelic richness expressing strong heterozygosity.

The work also addresses the phytosanitary situation of the vines in Kerkennah, identifying the main pests and the means of control. The diagnosis of the virological state of the vines also showed a low rate or even absence of infection which could be due to the limitation of the introduction of foreign plant material which could have contaminated the indigenous varieties of Kerkennah.

Finally, this work highlights the characteristics and importance of viticulture in the socio-economic life of the archipelago, as well as in the preservation of its cultural and traditional heritage. Such a rich heritage, unique and free from viral diseases, constitutes an asset for these islands, for their inhabitants but also a very useful support for vine selection and improvement programs. The protection, development and enhancement of this rich heritage are priorities, with a view to guaranteeing a sustainable future for the archipelago and its inhabitants.

ملخص

كتاب "التراث وتقاليد زراعة الكروم في قرقنة"، هو دراسة شاملة تسلط الضوء على تاريخ العنب وثقافة النبيذ في أرخبيل قرقنة التونسي، وهي بيئة خاصة وصعبة، تشهد على تاريخ حضارات البحر الأبيض المتوسط، والتي على الرغم من هشاشتها لا تزال المنطقة مفضلة للغاية.

يصف المؤلفون الكرمة في السياق التاريخي والبيئي للإقليم المحلي لقرقنة. الكرمة، هذا التراث الجيني القيم هو جزء لا يتجزأ، ليس فقط من نظام الإنتاج الزراعي ولكن أيضًا من التاريخ الغني للأرخبيل. على الرغم من القيود المناخية مثل الجفاف، وتدهور الغطاء النباتي، والتصحر، وملوحة التربة، وتغير المناخ، فإن الكرمة التي أدخلها القرطاجيون لا تزال قائمة، وتتبع أنماط تطور قطع الأراضي وطرق الإدارة المعمول بها.

يستكشف الكتاب بعد ذلك الأساليب والمعرفة التقليدية بالإضافة إلى ممارسات زراعة الكروم وإنتاج العنب الطازج والزبيب والنبيذ، مع إشارة خاصة إلى صنف العنب المحلي الرائد في الجزيرة : عسلي، المستخدم للاستهلاك الطازج، التجفيف والنبيذ الحرفي المحلي. كما تم وصف التكاثر والتوصيف الفسيولوجي لتطور الأوراق والبراعم لصنف العنب عسلي مع دراسة مفصلة لعنبه الطازج والمجفف والنبيذ واستخدامه في المطبخ القرقني. تتمتع الأنواع المختلفة المتحددة من كروم قرقنة بجودة استثنائية للفاكهة ومشتقاتها، ولكن أيضًا بالقدرة على التكيف والمرونة مع الجفاف الشديد المتزايد، وتتشكل هذه الأنواع مستودعات للجينات ذات الأهمية التي لا جدال فيها. في هذا العمل، تم وصف هذه الأنواع وتقييم تباينها عن طريق التحليلات الأمبلوجرافية والبيوكيميائية والخلوية والجزيئية. تم تكوين بطاقات للأنواع الثمانية التمثيلية لتشكيلة كروم قرقنة: عسلي، مهدوي، جربي، كحلي، مرصاوي، حمري، داليا وتونسي. تضاف إلى هذه البطاقات لوحات توضح الورقة، البراعم، العنقود، الزهرة والبذور، الأعضاء الأساسية للصنف. تم تنفيذ الوصف الأمبلوغرافي وفقًا لتدوين الاتحاد الدولي لحماية الأصناف النباتية الجديدة (UPOV)، المجلس الدولي للموارد الوراثية (IPGRI)، والمنظمة الدولية للكروم والنبيذ (OIV) محددة السمات الكمية والنوعية للأعضاء المختلفة لنبتة العنب في مراحل محددة من التطور الخضري.

و أظهر التوصيف الخلوي بتقنية التدفق الخلوي عدد الكروموسومات للأصناف الممثلة للكروم المحلية المزروعة في قرقنة و التي تشمل ثنائية الصبغيات (2n=38).

ومن جهتها كشفت التحاليل الجزيئية للتراكيب الوراثية المختلفة بواسطة الواسمات الجزيئية النووية لتكرار التسلسل البسيط، تنوعًا وراثيًا قويًا لأصناف الكروم التي تم تحليلها بالإضافة إلى تراء البدائل (الأليلات) الذي يعبر عن ارتفاع كبير في مستوى الأنواع الغير المتباينة وراثيا.

Chapter I

The Kerkennah archipelago

Kerkennah)<j^j^(sometimes spelled **Karkenah** , **Karkena** , **Kerkenna** or **Kerkena** , is an archipelago located in the heart of the Mediterranean about twenty kilometers from the Tunisian coast near the city of Sfax; Sfax being the second city of Tunisia after Tunis, it is also one of the most important commercial crossroads in the country.

Kerkennah is composed of the main ones : Gharbi (^£), also known as Mellita) and the name of the village near the village, and Chergui (^j^(or Grand Kerkennah and several other places: Gremdi)^^j3), Roumadia)^-'^JJ), Rakadia)4-^J(, Sefnou)>^(, Charmadia)^J-1^J^(, Ch'hima)4 ≙;≙ JI (-, Keblia)AJJ3 (, Jeblia)AJJJ^(, El Froukh)£JJ^(, Firkik)^£j*(, Belgharsa)<^jiJb(and El Haj Hmida) ⁵ 4^ C^ ¹) (Fig. 1).

The perimeter of the archipelago exceeds 160 kilometers. The Kerkennah Islands are located between 11° and 11° 20' east of the Greenwich meridian and between 34° 36' and 34° 50' north latitude, all extending over around thirty kilometers but covering only 150 km ² of surface area due to its narrowness: on these strips of land no point is more than 5 km from the shore, hence the importance of the sea in the natural landscape and its evolution in the life of the Kerkennians .

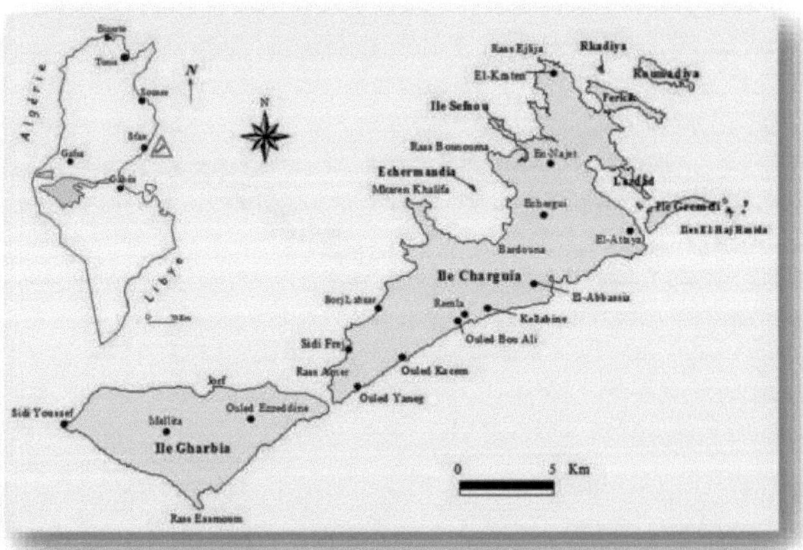

Figure 1. Location map of the Kerkennah Islands. Topographic sheets at 1/25000 (Fehri, 2011)

The topography of the archipelago is very low, barely emerging from a vast area of shoals in the Gulf of Gabes where the highest point is only 13 meters. This topography is remarkable for depressed and salinized expanses of chotts or sebkhas, today unsuitable for agriculture, alternating with more or less marked backs of land which correspond to Quaternary limestone crusts partially covered with aeolian sands and covering them -even red clay formations from the Mio-Pliocene that the sea cuts into cliffs on the west side (Fig. 2).

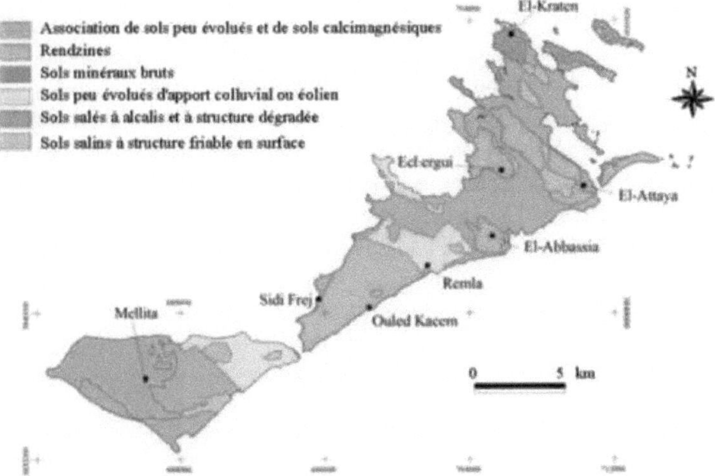

Figure 2. Pëdological map of the Kerkennah Islands (Ben Hassine and Besbes, 1994 cited by Fehri, 2011)

Migration towards the archipelago will then mark Kerkennah, attracting continentals from the Sahel, Sfax, Mahdia, Djerba, who will bring their techniques, exchange their traditions and draw inspiration from construction processes adapted to the winds to become sedentary. The resources of these islands have always been based on agricultural activities focused mainly on fishing, the cultivation of date palms, olive trees and vines.

Today, Kerkennah, with its eternal beauty identical to that described by Herodotus, witness to the history of Mediterranean civilizations, suffers from the absence of a sustainable development strategy.

The Kerkennah Islands are located on the north coast of the Gulf of Gabes which is itself a fragile, vulnerable physical environment, with limited resources, highly coveted and which experiences various forms of degradation which manifest themselves through the problems of continental erosion, marine erosion and land salinization as well as the extension of sebkhas (SPA/RAC - UN Environment/PAM, 2019). In fact, the sebkhas and chotts occupy a third of the total surface area of the islands.

Despite the constraints of the natural environment, agriculture in Kerkennah remains an important activity; it constitutes the second economic sector of the islands after fishing. It is above all a traditional subsistence agriculture, mainly oriented towards self-consumption and which must cope with the climatic and edaphic constraints of the natural environment. Agriculture is characterized by the predominance of irrigated

9

arboriculture including vines, fig trees (which are often combined), and olive trees with a reduced place for irrigated crops (Ben Salah, 2003). Furthermore, one of the particularities of the archipelago is the palm grove which represents a real source of wealth.

Fishing and agriculture, pillar sectors of the local economy, are going through a serious crisis generated in particular by the abandonment of traditional practices well adapted to the natural environment and the emigration of young people attracted by other activities outside the 'archipelago. To remedy this situation, the different actors at the local, regional and national levels have designed a large number of projects, often within the framework of international cooperation (Bouzid and Megdiche, 2006). Among their objectives: The conservation of biodiversity in the archipelago, the promotion of genetic heritage, the maintenance of traditional practices threatened with disappearance as well as the transmission of indigenous know-how relating to the conservation and rational use of biodiversity .

Agricultural land occupies only 3450 ha, or 22% of the total area of the Kerkennah archipelago (around 15,700 ha), while the rest of the area is covered by the Ghaba.

"غابة" (palm trees and forest trees (34%)), sebkhas (37%) and by the urban sector (7%) (Ben Salah, 2003). The area occupied by fruit trees in the archipelago is estimated at 923 ha (CRDA-Sfax, 2000). This area is far lower than that put forward by Andre Louis (1961) which was 6190 ha. This implies a significant reduction in surface area (5267 ha) which would have caused a considerable loss of genetic resources at the level of species and varieties.

Chapter II

Vineyard and environmental context of Kerkennah

A clear deterioration of the plant cover characterizes all the crops practiced on these islands explained by the rising water levels, desertification and salinization of the soil which constitute limiting factors which hinder the maintenance of crops. In this particular context, the cultivation of indigenous vines persists.

This culture owes its extent in these islands to the Carthaginians who introduced the cultivation of vines and olive trees, making the lands of Mellita and Chergui, today, vestiges of this ancient life with ocher, sandy, warm colors. and fragrant (Fig. 3).

Figure 3. Landscape of Kerkennah joining vines, palm trees, pomegranate trees and olive trees

Kerkennah has a semi-arid climate with very low rainfall, averaging 228 mm/year. Humidity levels are very high and temperatures, although they do not show strong variations on the monthly averages, can easily reach 40°C in the summer period. The archipelago is regularly swept by hot winds with more or less frequent episodes of sirocco.

La Hamada is the most famous area for growing vines; the soil is mainly clayey-sandy and very shallow. The vineyard production area is located on a limestone crust.

Kerkennah is until today renowned for its vines generally scattered in family orchards grown in high goblet with free arms, which allows the development of significant vegetation which shelters the clusters from the heat and the strong maritime winds (Harbi Ben Slimane, 1999).

In these vineyards, we find almost the same range of varieties as that of Sfax, with a predominance of the Asli variety used for fresh consumption, for drying and for local artisanal winemaking (Fig. 4).

Figure 4. Asli-Kerkennah cluster

A study carried out by Bassouls (2016) relating to the indigenous Asli variety in the restrictive pedoclimatic context of Kerkennah was based first of all on an identification of the cultural practices and the cultural weight of this grape variety among the traditions of the Kerkennians. It provides information on the one hand on the identification of the different production sites and on the other hand on the characterization of the pedoclimatic context of the identified area. The results of the preliminary investigation carried out by Ms Bassouls and directed by Mr Habib Ben Cheikha, Director of the Territorial Popularization Unit of Kerkennah, highlight certain particularities of the cultivation of vines, mainly Asli, on the islands. These data made it possible to understand the relationship of the Kerkenniens with this particular grape variety and to highlight the patterns of evolution of the plots and the practical management methods (Fig. 5).

Figure 5. Mapping of the four Asli production zones in the Kerkennah Islands (Bassouls, 2016)

Regarding the land situation on the islands, the results of the survey highlighted disparities in the means of obtaining land depending on the municipalities (Fig. 6).

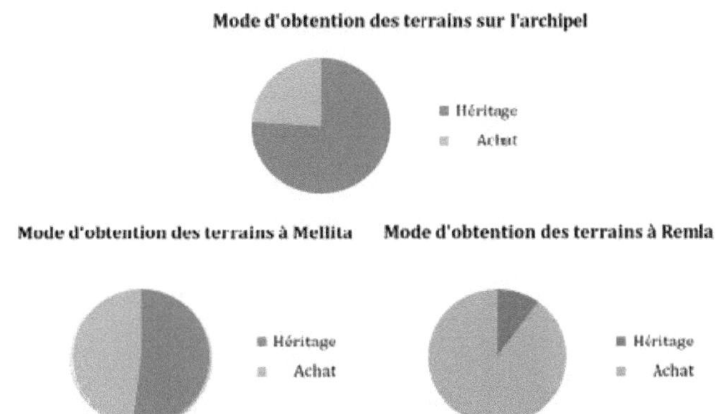

Figure 6. Method of obtaining land on the archipelago and by municipality (Bassouls, 2016)
The graphs from the survey show that the main means of obtaining land on the archipelago remains inheritance. These disparities should be related to the map of the irrigated areas and boreholes of the archipelago (Fig. 7).

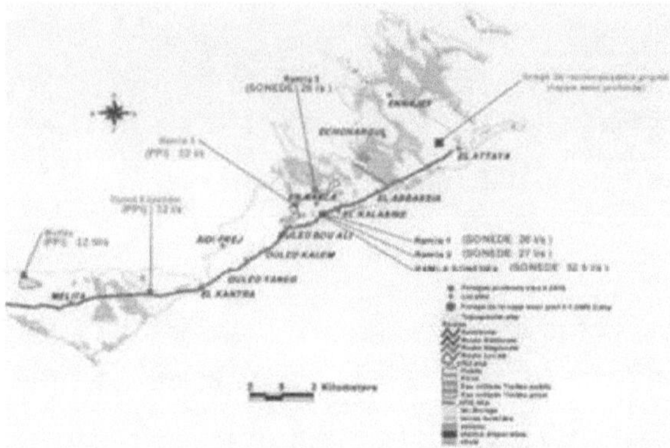

Figure 7. Map of irrigated and drilling perimeters on the archipelago (CRDA-Sfax, 2016)

However, in certain irrigated areas such as Mellita, land has gained market value and there is an upsurge in the purchase of plots in these areas.

If we put these observations in relation to the graph of cultivated areas, we notice that in the communes where the main means of obtaining land is inheritance, namely Remla (or Ramla) and Jouaber, the cultivated areas are very small (between 0.04 ha and 1 ha). While in the communes of Mellita and Ouled Ezzedine, where we note the presence of irrigated perimeters, the areas cultivated by wine growers are larger (between 0.5 and 5 ha).

The situation reveals very small plots in the majority of cases, especially in Remla. Access to land, mainly through inheritance, leads to land fragmentation, except in areas which gain market value through the establishment of irrigated perimeters (Fig. 8).

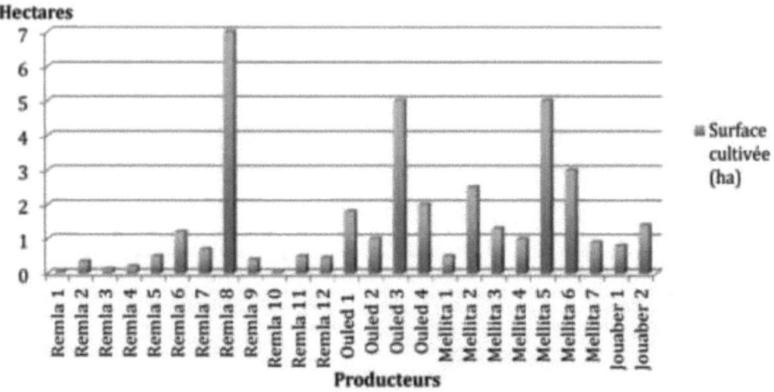

Figure 8. Cultivated areas in the municipalities of Kerkennah (Bassouls, 2016)

The study of crops and their associations in the plots reveals a very large proportion of associated crops; out of the 25 farmers surveyed, 2 wine growers cultivate vines in monoculture.

In the areas of Remla and Jouaber, the main association is vine-fig, some have introduced the olive tree with drip irrigation.

While in Mellita and Ouled Ezzedine, we observe a diversification of associations with the introduction of pomegranate and almond trees and a greater proportion of olive trees (Fig. 9).

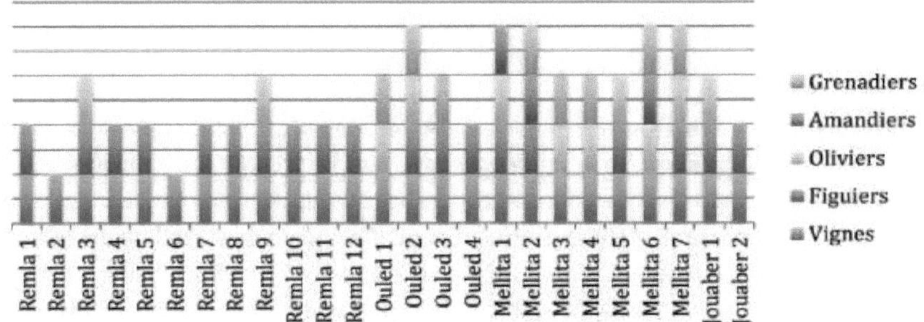

Figure 9. Cultures and their associations in Kerkennah (Bassouls, 2016)

In addition to the diversification of crops in associations in the plots, we observe greater variability in the grape varieties cultivated in the areas of Mellita, Ouled Ezzedine and Jouaber. Indeed, in these communes, some winegrowers cultivate up to 6 grape varieties on the same plot. On the contrary, in the Remla area, the main and practically the only grape variety cultivated is Asli. According to winegrowers' testimonies, the Remla area is the ancestral production area of this grape variety, from which it was dispersed to other areas of the archipelago (Fig. 10).

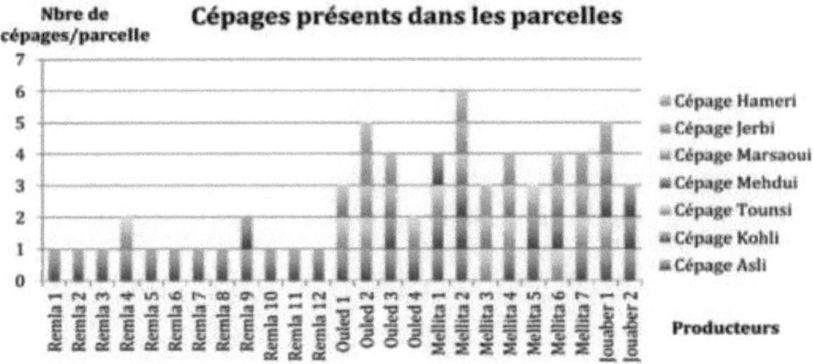

Figure 10. Distribution of grape varieties in the different Kerkenian communes (Bassouls, 2016)

From the point of view of cultivation techniques and more particularly irrigation, there is great variability between plots. This variability is directly linked to proximity to an irrigated perimeter. We note that the main production area of Asli (Remla) is rainfed (Fig. 11). However, we note that the producers of this area, able to practice irrigation, have diversified their income by the introduction of olive tree cultivation.

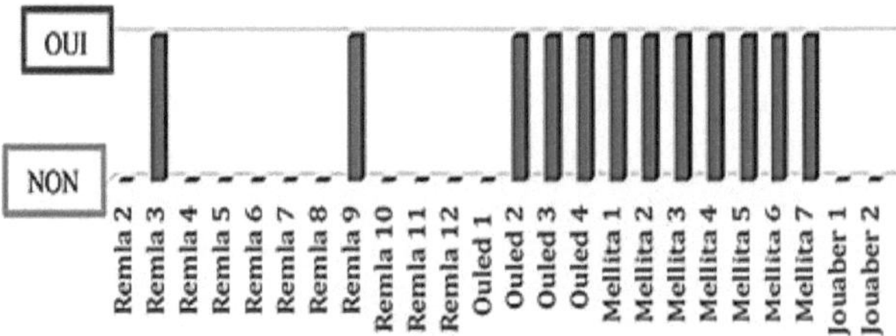

Figure 11. Irrigation practice according to the municipality in Kerkennah (Bassouls, 2016)

The production of Asli from Remla, identified as the main production area of this grape variety, is used for different purposes.

The first conclusions resulting from this analysis indicate, on the one hand, the existence of an ancestral area of cultivation of the indigenous Asli grape variety, around Remla. Asli plants are grown dry and in association with fig trees on very small plots. On the other hand, the appearance of irrigated perimeters has modified the agricultural landscape of the islands by favoring the introduction of new crops such as the olive tree, which tends to take precedence over traditional crops.

The different production zones of the Asli grape variety, identified by reading the landscape and the testimonies of the different stakeholders, are presented in figure 12.

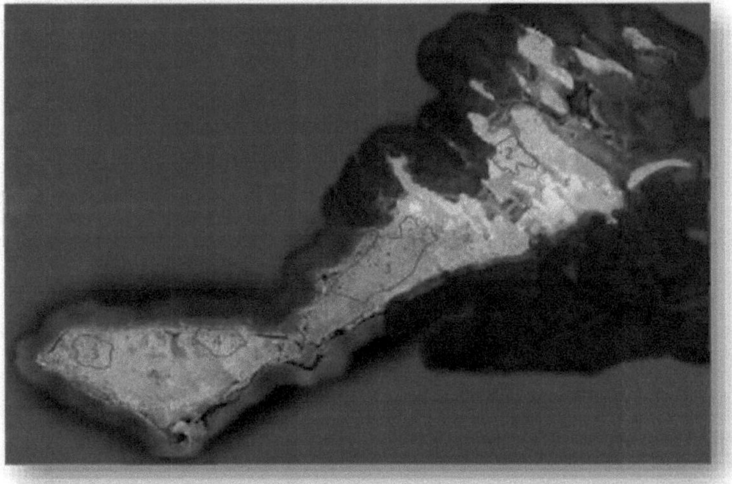

Figure 12. Mapping of Asli grape production areas. 1. Remla; 2. Mellita; 3. Ouled Ezzedine; 4. Jouaber (Bassouls, 2016)

This mapping shows that the largest production area corresponds to zone 1. This was confirmed by the field investigation.

Figure 13. Mapping of the main Asli production area (Bassouls, 2016)

This area, belonging to the commune of Remla, is located on a particular geological entity, called Hamada, which is a rocky limestone plateau, elevated from desert areas such as the Sahara. However, the Hamada of Remla does not have much of a plateau when we look at its altitude which is 2 to 3 m above sea level. This area of almost 400 ha surrounds the village of Remla and extends practically to the coasts on either side of the island. There are mainly small plots not exceeding one hectare planted with vines and fig trees in association.

On the eastern part of the Hamada, there is today an irrigated area, the vineyard plantations have been replaced by olive trees. Many palm trees are still present between the plots and even inside some. However, as these are no longer maintained, the palm grove is quickly dwindling. Beyond this zone, to the East, extends a Sebkah which tends to gain ground day by day. It is in this Hamada that Asli vines were traditionally cultivated. According to the winegrowers of Kerkennah, some of the vines that can be found there are more than 250 years old, and something particularly rare is that the cultivation is still carried out in franc de pied.

A few years ago, a CTV initiative aimed to produce grafted feet but this project did not last. The only feet resulting from this initiative are today on CTV land. Nevertheless, the taste qualities of Asli franc de pied must be preserved; for this reason, grafting is not recommended as it would cause a change in the quality of the production.

We are currently in the presence of an ancestral cultivation area threatened by the introduction of more profitable crops such as the olive tree, thanks to new management methods such as irrigation and the introduction of chemical inputs.

Chapter III

Vine and traditions in Kerkennah

Allemand-Martin (1907) recounted the vineyards of Kerkennah and the traditions of the inhabitants in the use of grapes, whether for consumption or processing (Rhouma *et al.,* 2005). Nowadays, the harvest is celebrated in Kerkennah, a custom practiced for centuries. For this, a ceremony is announced by the tambourine in villages renowned for their vine cultivation such as the Q'baylia (^JJJ/JS) in Ramla to announce the day "d" of the grape picking, an event marked by the departure of well-dressed women wearing their typical Kerkennah scarves (Fig. 14).

Figure 14. Old scarf with grape cluster motifs

A bright red floral scarf with grape cluster motifs all around.

Today, only one specimen remains at the Island Heritage Museum in Abassia.

Groups of women, on this festive day, head towards the orchards at dawn, humming ancient songs, it is the period of "guassan el aneb" (^j*Jl j'-^) that the Kerkennians are waiting for to start the harvest.

3.1. The Asli variety from Kerkennah

The Asli variety (J"^), meaning taste of honey, is considered the star variety of the island (Fig. 4). It is the most widespread and also the oldest since, according to certain inhabitants, it exists today Today, Asli plants are over 250 years old and are grown on their own mainly for the production of dried grapes. In fact, only a small part of the grape variety's production (less than 5%) is consumed fresh by the farmers. The Asli grape variety also has characteristics of a vat variety, which allows the excess production to be virnfied by raisining.

As the fruits approach maturity, birds begin to cause damage to the grapes from sunrise and late in the day, starting with the earliest fruits. In order to protect the vine crops from birds, the Kerkennians frequently use scraps of worn fishing nets collected from fishermen to cover the entire vine plant and thus minimize damage to the grape clusters (Fig. 15).

Figure 15. Vine plant entirely covered with fishing nets acting as an anti-sparrow net

In addition to its effectiveness, this economic and ecological strategy testifies to the intelligence of the Kerkennian who lives in harmony with his environment.

3.2. Fresh grapes from the Asli grape variety

During a study carried out in 2015-2019 by researchers from the National Institute of Agronomic Research of Tunisia (INRAT) on the varietal behavior of the Asli grape variety subjected to irrigation and its effect on the morphometric and chemical quality of the grapes fresh, bunches from the Asli grape variety were analyzed.

The study concerned two plots cultivated, one irrigated (drip) and the other rainfed, located in the center of the Kerkennah archipelago (La Ramla - Fig. 12). Localized irrigation is done from a surface well with water loaded with 3.6 to 6 g/l of salt which varies depending on the season. Clusters of the Asli grape variety were picked at full maturity (early August) from plants chosen randomly in the vineyard. Ten clusters per treatment (irrigated or rainfed) were analyzed. Morphometric analyzes focused on the length, width, weight of the cluster and berries as well as the number of berries per cluster. The firmness of the grapes was measured on two opposite sides of the berry using a benchtop penetrometer (Fruit Texture Analyzer, GUSS Manufacturing, South Africa) equipped with an 11 mm diameter tip. The shape and color of the berry were determined using vine descriptors (IPGRI, UPOV, OIV, 1997).

The soluble sugar content (TSS), expressed as a percentage and representing the grams of sugars per 100 ml of juice, was determined using a digital refractometer (Type OPTECH GmbH, Munchen, Germany). The titratable acidity (TA) was determined by titrimetric dosage of the juice with NaOH soda (0.1 N). The juice was obtained by manually pressing the grape berries on a sieve to eliminate the skin and seeds each time. In this way, it was possible to determine the number of seeds per berry.

The morphometric and biochemical analyzes of Asli carried out are summarized in Figure 16. The irrigated plants produced clusters measuring on average 18.3 cm in length and weighing more than 202 g/cluster for a total of 191 berries per cluster. In rainfed conditions, the cluster reached an average length of 16.9 cm and an average weight of 217 g/cluster for a total of 156 berries per cluster. The weight per berry is greater in the rainfed type (1.82 g). Analysis of the juice revealed a large difference in the soluble sugar content (TSS) varying from 23.5% in irrigated plants to 26.6% in rainfed plants. Titratable acidity (TA), expressed in relation to the tartaric acid content, showed values of 1.26 and 1.01 g/l respectively in grapes grown from irrigated plants and those grown in rainfed conditions. The TSS/AT ratio, referring to the maturity index, showed a higher level of maturity in plants grown in rainfed conditions.

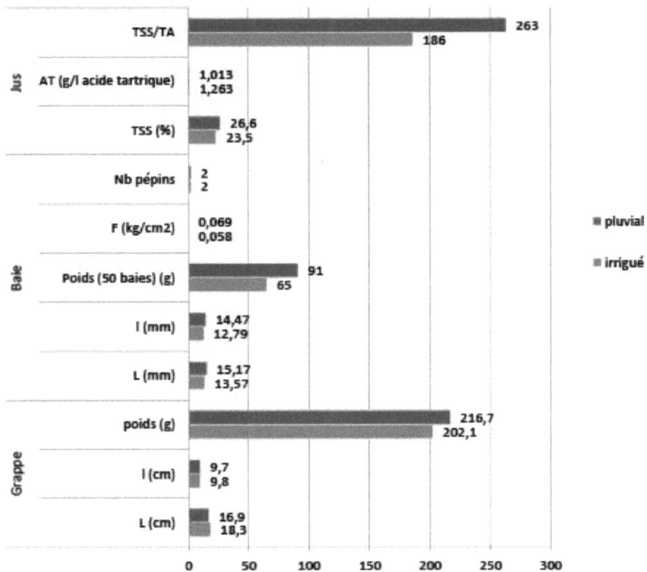

Figure 16. Effect of irrigation on the morphochemical parameters of Asli grapes (L: length, l: width, F: firmness, TSS: soluble sugar content, TA: titratable acidity)

Irrigation has probably not improved the qualitative performance of this grape variety; on the contrary, the soluble sugar content and the maturity index remain more interesting in vines grown under rainfed conditions. However, the long periods of drought and lack of water experienced in recent years require the use of irrigation to maintain a minimum of growth and production.

As for the quality of the clusters, measurements showed that irrigated plants produced larger clusters with an average length of 18.3 cm and a width of 9.8 cm. Whereas in rainfed conditions, the clusters reached an average length of 16.9 cm and a width of 9.7 cm. However, the weight of the cluster is greater in rainfed vines (217 g) due to the presence of larger berries. Indeed, the weight of 50 berries showed a large difference under the effect of irrigation in favor of plants grown under rainfed conditions.

Under irrigation, the weight of 50 berries is 65 g, compared to 91 g under rainfed conditions. Irrigation seems to have a recalcitrant effect on fruit size; irrigated plants produce smaller berries. This can be explained by the quality of the water used for irrigation. The Kerkennah area is known for its relatively salt-laden water, but above all for its poor, poorly aerated and shallow soils (Louis, 1963) which can lead to salinization of arable land.

The Asli grape variety grown dry or irrigated produced berries measuring on average 15.2 mm in length with 14.5 mm in width and 13.6 mm in length with 12.8 mm in width, respectively.

It should also be noted that irrigation reduced the firmness of the grapes. Under irrigation, fruit firmness showed an average value of 0.058 kg/cm2 compared to 0.069 kg/cm2 in non-irrigated plants. Thus, grapes grown under rainwater conditions are remarkably firmer and larger in size (Fig. 17). The number of seeds per berry being the same in both irrigated and non-irrigated vines.

B

Figure 17. Clusters of Asli grown in Kerkennah **A.** grape variety grown dry; **B.** cëpage conducted in irrigtic

Analysis of the Asli grape juice showed a high level of soluble sugars (TSS). Indeed, grapes, compared to other fruits produced in the archipelago, have the highest soluble sugar contents (21% to 24%) (Zemni, 2007) to be compared with figs (16.1% to 18.4%). (Trad *et al.* 2012) or apricots (11.1% to 14.1%) (Lachkar, 2014).

Asli stands out compared to other grape varieties grown in the Kerkennah region. A previous study carried out by Zemni (2007) revealed that Asli would be the sweetest grape variety on the island showing values of around 24% in comparison with Mehdoui and Kahli having respectively 21.5% and 20% soluble sugars.

As for our study, driving style also influenced the soluble sugar content. Indeed, the latter was higher in grapes grown under rainwater (26.6%) compared to grapes grown under irrigation (23.5%). This variation contributes to a slight modification in the perception of the taste and organoleptic quality of the fruit.

On the contrary, titratable acidity (TA), expressed in grams of tartaric acid per liter of juice, commonly used in viticulture and renology to judge the acidity of the harvest (Boulton, 1980), was lower in grapes grown in rainfed conditions. This parameter varies depending on soil, climate, variety and irrigation (Boulton, 1980).

Another important parameter in determining fruit quality, the TSS/AT ratio or maturity index (Karacali, 2009), being higher in grapes produced in rainfed conditions. This leads us to the conclusion that irrigation delays the ripening of Asli grapes or that water stress anticipates fruit ripening. This was confirmed by the results published by Van Leeuwen and Vivin (2008) which showed that moderate water stress increases the speed of maturation because of less competition for carbon (growth arrest). Water stress is thus considered an important factor in precocity. Indeed, the development of the vine and the maturation of the grapes are strongly influenced by the water regime. The latter depends both on soil water reserves, climatic parameters (precipitation, ETP) and the architecture of the vegetation (Van Leeuwen and Vivin, 2008).

In the Asli grape variety, irrigation did not improve the quality of the grapes. On the contrary, the size of the berries, the soluble sugar content and the maturity index were better in the rainfed vines, attesting to the very good adaptation of the Asli grape variety to the arid conditions which mark the Kerkennah archipelago. In this context, studies carried out by Ben Salem *et al.* (2000) showed the strong adaptation of several local cultivars, among them Asli, established under extreme conditions of temperature and light in the south of Tunisia, having demonstrated exceptional vigor and early production.

In conclusion, it should be noted that the qualitative performances of the Asli grape variety proved to be better in rainfed mode. However, changing climatic conditions in the Kerkennah archipelago require recourse to irrigation to maintain a minimum level of growth and production. Viticulture, as traditional as it is, has always contributed to the food security of the Kerkenian populations.

The recovery of rainwater (design of cisterns) and the improvement of soil fertility through organic amendments could constitute possible solutions for successful cultivation in marginal lands. In these particularly dry conditions, irrigation must above all be a factor in increasing yield. By maintaining a moderate water deficit, it can also be a factor of quality.

3.3. The raisin of the Asli grape variety: The Zebib of Asli

Minangouin (1901) impressed by the Asli grape variety present for a long time in the Kerkennah vineyard said: *"Asli is a cëpaдe appɀëaë for the finesse of its fruits, its very sweet taste and its pɀëcocИë , but its grain is very small and its skin little ix'sislanle"* . He concluded that it is very suitable for dried grapes. He evaluated its production at 40 kg of raisins from 100 kg of fresh Asli grapes. Likewise, thanks to their drying quality, Asli grapes have been highly recommended for the production of raisins (Ghrairi *et al.* 2013).

The dried grape, in Arabic Zebib (^?j), constitutes for humans a food with an excellent energy supply, available all year round, easy to take and easy to store. It is also very appreciated by consumers in all regions of the world thanks to its taste qualities (sweet flavor, perfumes, etc.).

The consumption of dried grapes by humans dates back to prehistoric times. Early human gatherers and hunters probably recognized the qualities of wild grapes and noticed the transformation of the grapes into a dried edible form after falling from the vine plant and lying in the sun. In the Neolithic, grapes were probably dried for storage and travel. Archaeological excavations in the Mediterranean region (Lachish in Palestine) have revealed prehistoric wall paintings dating back to the Bronze Age proving a very ancient use of raisins in food and decorations.

The early Phoenicians and Egyptians are said to have popularized the production of raisins and their use throughout the world, where they were valued for their easy storage and transportation (Williamson and Carughi, 2010). Worldwide, dried grapes are classified among the most consumed dried fruits and the most used grape varieties are represented essentially by Sultanine, Golden, Muscata and Black Corinth (Ferradji *et al.* 2008).

In Tunisia, the local varieties which are best prepared for drying and which produce raisins of exceptional quality are Meski Raf-Raf (Harbi Ben Slimane, 2005) and Asli (from Sfax and Kerkennah) (Harbi Ben Slimane *et al.* 2014;

In the Kerkennah Islands, production is self-consumed in the majority of cases, after transformation into Zebib. The dishes and pastries made from this ingredient are an integral part of the traditions and customs of the Kerkennah Islands.

3.3.1. Transformation of Asli grapes into Zebib

At Madame Fathila Azzabou living in Ramla, the Asli grapes are generally intended entirely for drying to make Zebib, as is the case with many Kerkennians: The grapes are picked by hand (Fig. 18).

Figure 18. Grape picking

The bunches are cut, placed in airy baskets carried in large esparto bags (^UJ^Jl) designated by el alegua (<^*Jl) or errsala (Huijil), and transported on the backs of donkeys to the place drying (Fig. 19).

Figure 19. Transport of grapes to drying places

The bunches are dried either directly on the sand (Fig. 20), or on a layer of plant called El Gazej (^lj3Jl) which is *Pituranthos scoparius* , previously spread on the sand.

Figure 20. Drying *Zebib*

Two to three weeks are necessary to dry the grapes (Fig. 21); they will then be detached from the stalks and pedicels.

At the time of conservation, the Zebib is washed with sea water as a natural means of disinfection. The raisins are then packed into the Zirs (j^j) (Fig. 22) or the Khabias (ЧД -) (Zirs has two handles instead of four).

The Zir or the Khabia, which can be made of raw or enameled earth, allow the raisins to be preserved for up to a year.

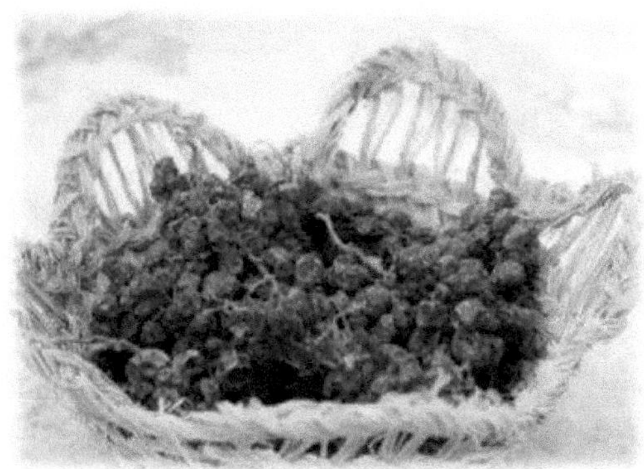

Figure 21. Basket of raisins

Figure 22. Zir for preserving raisins

3.3.2. The nutritional value of raisins from the Asli grape variety

The consumption of raisins offers numerous benefits for human health that contribute both to the promotion of general well-being and to the prevention of many chronic diseases, such as gastrointestinal diseases, cardiovascular diseases, type diabetes. 2 (Anderson and Waters, 2013) as well as dental caries (Schuster *et al.* 2017).

Dried grapes are approximately 4 times more caloric than fresh grapes: 100 g can provide approximately 321 kcal, or 1360 KJ (Anses, 2022). In addition to their richness in carbohydrates, raisins contain approximately 4 times more minerals and trace elements and a low water content (Tab. 1).

Table 1. Nutritional composition of raisins (Anses, 2022)

Raisin (Value per 100 g)	
Energetic value : 321 kcal	**Water** : 9.9 - 22.5 g
Carbohydrates : 73.2 g	**Fiber** : 2.8 g
Proteins : 2.07 - 3.8 g	**Organic acids** : 4.7 g
Lipids : 0,1 - 1.6 g	**Cholesterol** : 0 g

Dried grapes are also interesting for their contribution in dietary fiber. An analysis of the dietary fiber composition showed a predominance of pectins, polysaccharides as well as mannose and glucose residues.

High levels of fructans (6 to 8%) were also found in samples of sun-dried or artificially dehydrated raisins (Schuster *et al.* 2017). Fructans act as prebiotics that stimulate beneficial intestinal microflora (bifidobacterium and lactobacilli) .

Several studies suggest that fructans may protect against colorectal cancer and reduce blood triglycerides (Bell, 2011).

Analysis of the nutritional value of Zebib from the Asli de Kerkennah variety (Fig. 23) revealed a richness in mineral elements, particularly potassium (473 mg/100 g), calcium (187 mg/100 g) and magnesium (32 mg/100 g), in addition to a high sodium content (139.5 mg/100 g) which would be closely linked to the salinity of the waters of the archipelago (Tab. 2).

Figure 23. Asli raisins

Table 2. Mineral element contents of Asli Kerkennah raisins compared to the Ciqual reference table

Mineral elements	Raisin from Asli Kerkennah (Value per 100 g)	Ciqual reference table for dried grapes (Value per 100 g)
Copper (Cu)	0.196 mg	0.23 - 0.5 mg
Iron (Fe)	1,954 mg	1.13 - 5.2 mg
Zinc (Zn)	0.186 mg	0.12 - 0.52 mg
Sodium (Na)	139.435 mg	5 - 39 - > mg
Potassium (K)	472.77 mg	584 - 862 mg
Phosphore (P)	63.502 mg	74 - 125 mg
Manganese (Mn)	0.219 mg	0.22 - 0.35 mg
Magnesium (Mg)	31.941 mg	27 - 38 mg
Calcium (Ca)	186.999 mg	28 - 77 mg

La table de reference Ciqual (Anses, 2022)

Dried grapes also contain a small amount of boron (2.2 mg/100 g) but with significant effects. According to the World Health Organization, the daily intake of boron for adults is 1 to 13 mg/day (WHO, 1996). Boron participates in the metabolism of calcium, copper , magnesium , amino acids (constituents of proteins), glucose (sugar circulating in the body), triglycerides (fats) and estrogens. It seems to intervene in erythropoiesis (the formation of red blood cells), immune defenses and cerebral functioning, and would have an anti-inflammatory action, but its best documented effects concern its positive impact on bone: calcification and stabilization of bone. bone mass (Martin, 2001; EFSA, 2004). Thus, the simultaneous intake of calcium and boron by raisins would be very important for the growth of the skeleton and teeth in children (EFSA, 2004) and the prevention of osteoporosis (Nielsen, 1996) and arthritis in adults (Naghii and Samman, 1993). In addition, Zebib from Asli de Kerkennah turns out to be very rich in vitamin E and its derivatives: tocopherols and tocotrienols (Tab. 3).

The value recorded for vitamin E is 0.4 mg/100 g, more than double the maximum value given by the international reference table Ciqual (Anses, 2022).

Table 3. Vitamin contents of Zebib from Asli Kerkennah

Vitamins	Raisin from Asli Kerkennah (Value per 100 g)
Alpha-Tocopherol (Vitamin E)	0.4 mg
Beta-Tocopherol	8.71 mg
Gamma-Tocopherol	0.36 mg
Gamma-Tocotrienol	4.7 mg

Furthermore, dried grapes contain several phenolic compounds. These compounds influence sensory properties such as color, mouthfeel characteristics, taste and antioxidant characteristics. The polyphenol contents and antioxidant levels observed in raisins are among the highest compared to other fruits (Karadeniz *et al.* 2000; Jeszka-Skowron *et al.* 2017).

Indeed, raisins contain high concentrations of phenolic compounds, such as phenolic acids (gallic acids (316-1141 mg of gallic acid/100 g of raisins), coumaric, transcaftaric, trans-coutaric and ferulic), flavan-3-ols (catechin and epicatechin), flavonols (myricetin, quercetin and kaempferol) and anthocyanins (malvidin -3-O-glucosides and its acyl esters) (Karadeniz *et al.* 2000; Williamson *et al.* 2010; Murphy, 2012). Black raisins would be a good food source rich in anthocyanins responsible for their red, purple and blue color.

Linoleic acid (LA), belonging to the omega-6 family, has the particularity of being the only essential fatty acid that cannot be synthesized by the human body, hence the importance of its contribution by the diet (Kaur *et al.* 2014). The determination of fatty acids in Asli Kerkennah raisins (Tab. 4) revealed the predominance of linoleic acid (27%) and its main derivative dihomo-gamma-linolenic acid (18%). The human body converts LA into other fatty acids as needed, such as conjugated linoleic acid (CLA), which has anti-cancer activity.

Table 4. Fatty acid contents of Zebib from Asli Kerkennah

Fatty acids	Raisin from Asli Kerkennah	
Lauric acid	9.28%	
Palmitic acid	15.93%	
Palmitoleic acid		2.04%
Stearic acid		3.23%
Oleic acid		9.4%
Linoleic acid	26.89%	
Linolenic acid		2.9%
1 1-cicosdnoic acid	5.88%	
Dihomo-gamma-linolenic acid	17.83%	
Lignoceric acid	6.53%	

CLA intake via diet also leads to reduced atherosclerosis and reduced fatty streaks (Tony and Alphine, 2020).

Trials conducted on rabbits fed CLA revealed lower plasma triglycerides, lower plasma LDL cholesterol, and fewer aortic fatty lesions (Lerch *et al.* 2012). In addition, gamma linolenic acid participates in the production of human cells, in the proper functioning of the immune system and helps reduce the risk of cardiovascular diseases, the leading cause of mortality in the world, according to the WHO (Schuster *et al.* 2017).).

3.4. Assir or wine from the Asli grape variety

Speaking of the vines of Kerkennah and the winemaking traditions of the inhabitants, Allemand Martin (1907, 1940), said: " *The inhabitants make an excellent wine from the grapes annually, which can be юppzo ^ ë Marsala or Madeira, they could certainly obtain a very good yield and a wine priced at ë ^ ë* ". Minaingouin (1905) also indicated that the Asli of Kerkennah is used to make a wine very rich in alcohol (16 at 17°), reminiscent of Madeira wine. It is also a grape variety recommended for the production of liqueur wines. According to the Kerkennians, Asli, which has wine grape characteristics, was vinified by raisining over the years. where production was abundant. This process ensured longer conservation than the drying of the grapes itself, which is considered a currency for the islanders. Currently, this traditional winemaking process is increasingly rare and limited to it. production of small quantities of wine intended for family consumption As they are harvested, the clusters of Asli reserved for Assir (عصير) are spread on a layer of El Gazej The bunches of grapes attached to the stalk fill them. three quarters of a large container called Mahbes (^ ч ^) in terracotta with an average diameter of 70 cm and an average height of 40 cm (Fig. 24).

Figure 24. Mahbes[i] (محبس) terracotta

Figure 25. Grape juice extraction **Figure 26** . Recovery of grape juice by crushing

Figure 25. Extraction of grape juice by crushing

The extraction of grape juice is done by delicate crushing using carefully washed and well-dried feet. Subsequently, the collected grape juice (Fig. 26) is filtered into a large bucket using a sieve covered with a very fine mesh cloth (Fig. 27).

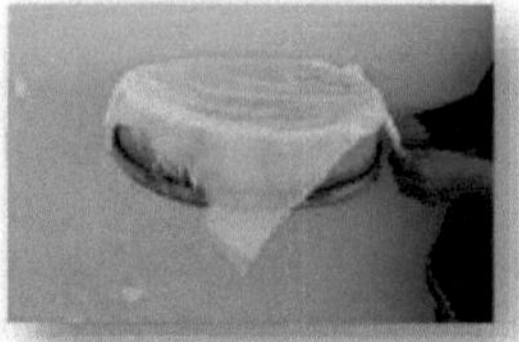

Figure 27. Sieve covered with very fine mesh fabric

terracotta jars or Khabias (ЧД -) which are then well cemented with clay (Fig. 28). The Khabias are kept for a minimum period of four months. With this technique, it is estimated that 20 liters of wine will be produced for a quantity of 70 kg of fresh Asli grapes (Fig. 29, 30).

Figure 28. Processing into wine in khabia **Figure 29.** Wine obtained from Asli

Figure 30. Storage of Assir in cellars and under beds. Abassia Museum (Fehri, 2014)

Asli winemaking is carried out by the families who own the largest vineyards. The wine, part of which is intended for family consumption, was well marketed and constituted an estimable source of income.

The typical Assir of Kerkennah is an organic wine without the addition of chemicals for preservation. This wine is consumed 5 to 6 months after the harvest.

Currently, we find less and less Assir following the decline in the cultivation of the grape variety. This decline is generally attributed to very old, increasingly fragmented vineyards as well as to the emigration of Kerkennians which dates from the 1960s. A situation accentuated by the severity of the climatic conditions of the archipelago, very influenced by global warming, lack of rainfall and increased soil salinity.

It is encouraging to know that the preservation of Asli grape cultivation is envisaged through measures such as the rejuvenation of plantations and the multiplication of orchards. These actions can have a significant impact on the revival of Asli grape production and the preservation of traditional winemaking processes.

The analysis of a sample of Assir (2016 production), carried out according to the Analysis Standards of the GIFRUITS-Tunisia Laboratory, revealed the following characteristics:

- *A high alcoholic degree of 13°30,*
- *Low total acidity, around 3.90*
- *An acceptable volatile acidity of 0.68*
- *A very low SO2 content not exceeding 11 mg/liter which indicates winemaking without SO2 additives.*
- *A density of 1021/ 20° which allows the designation of this wine as **a sweet wine** .*

1.5. The Asli grape in Kerkenian cuisine

The Asli grape is anchored in Kerkennian gastronomic heritage. Several of the recipes of Mrs. Rachida Ben Ezzeddine, whose biography will be recounted later, deserve to be mentioned in this work; recipes that she dictated to us with all the desire to safeguard these culinary practices to which she is extremely attached.

1.5.1. Charmoulet el Karnit

It is a dish generally presented in the meals of major ceremonies such as those of circumcisions or on the day of the feast of Aid es-Seghir.

Figure 31. Charmoulet el Karnit

It is a sauce made from onions cut into strips, browned in oil to which octopus tentacles cut into 3 to 4 cm pieces and raisins are added. For 2 Kg of raisins we use 3 Kg of onions. Add coriander, caraway, chili powder. Cook everything, remove the pieces of octopus (Karnit) as soon as they are cooked and sift the rest of the preparation. Return to low heat, adding the octopus pieces to the onion sauce.

1.5.2. Asli Vinegar

The residues from the winemaking technique consisting of the stems, skins and seeds of the grapes are used for the production of "a homemade vinegar".

Figure 32. Asli Vinegar

The technique consists of adding rainwater to the residues from the vinification of Asli. The preparation is kept for 20 days in the shade then filtered and distributed in tightly closed jars. This vinegar has an acetimetry degree of 4°56 without alcohol.

1.5.3. The Robb of Asli

The Asli berries are detached from their pedicels, washed in water and passed through a fine mesh sieve: ghorbel (J^J^). The fresh grapes are placed in a container on the fire, after a few minutes the heat is reduced and the preparation is passed again through the fine mesh sieve.

The mixture is then brought to a boil, then left to simmer over low heat for almost 2 hours.

Robb de Asli is often eaten in the morning for breakfast accompanied by hot fritters or ftaier (.лЛЦ) .

Figure 33. The Robb of Asli

1.5.4. Asli grape jam

The Asli grapes, whole or cut lengthwise, are placed in a thick-bottomed container with a little water.

Figure 34. Asli Grape Jam

For every three measures of grapes, add one measure of sugar. Everything is brought to the boil; the grapes which are floating above are removed and cooking is continued until the preparation solidifies.

In the meantime, the bottles to be used for preserving jam are sterilized in boiling water then dried on a clean cloth. They are then filled while leaving nearly 2 cm of empty space in each bottle. The bottle is then closed and inverted, lid down, until the contents have cooled. The jam thus obtained can be stored for up to 2 months in the open air.

1.5.5. Asli raisin laklouka

Laklouka is an ancient pastry from the region of Kerkennah and Sfax made from dried grapes of the Asli variety. To do this, 2,250 kg of raisins are washed, placed in a pot and covered with water. Everything is brought to a boil over medium heat until the Zebib grains swell.

The preparation is then crushed and then filtered through a fine mesh sieve.

The filtrate is then placed on the fire until the water reduces. 250 ml of olive oil is added and everything is put back on the heat until it has the consistency of a syrup. Then add 500 g of previously browned flour and continue cooking.

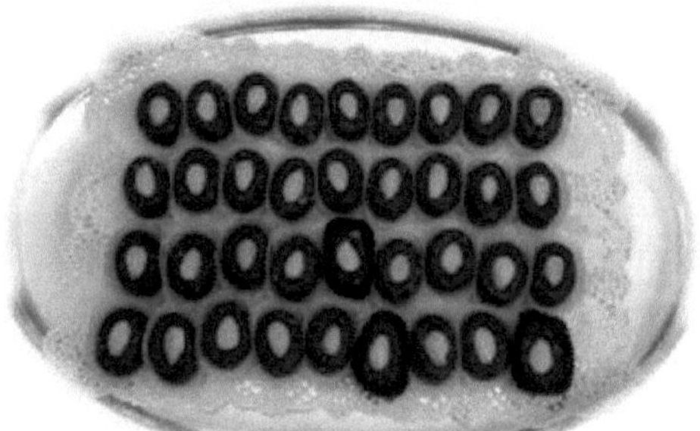

In the meantime, add 500 g of bsissa called Houar (J'J^) prepared from barley (J«-"), chickpea (^^), cumin (JJ-£), fennel (^ Ц ^), bark pomegranates (jj JJ^), coriander (JJIJ), anise (Sp^ AJA.), wheat (^—3), sesame seeds (j^P^), almonds (JJJ) and hazelnuts (*jj?) grilled and crushed.

We continue cooking with a large wooden spoon called medlek (<^b*).

Add the hot olive oil and knead again until the oil is completely integrated into the preparation.

After cooling, we shape small sticks to decorate with a molded almond browned in oil. Laklouka thus prepared can be kept for up to two years and its consumption must be moderate because it is very energetic.

A variation of the laklouka is the "Laklouket el Welada (S-^ JJI ASJKJ)". This laklouka is generally prepared for women the first days after childbirth. The preparation is based on dried grapes crushed with water and filtered, everything is placed over medium heat until the preparation browns.

We then add a little salt, flour previously sifted and browned, everything is then watered with carob water (^JJ^).

To the preparation are added dried figs *(^J*J£ A^JJ^)*, raisins and a bag filled with thyme (J^J), fenugreek (ЯД *) and fennel (^>b"u). The preparation should be consumed hot.

1.5.6. Sailors' couscous

It is a typical couscous from Kerkennah, sailors' couscous (SJW' ^^^^), prepared on the day of the installation of a traditional and organic fishing technique called Charfya (Vp (Fig. 36).

Charfya being a fishing technique typical of the archipelago where the Kerkennians share marine properties which are demarcated using palm leaves, composed of a large "big house" compartment, a small "small" compartment house", a lamp, two traps, a catching chamber and wings. The Charfya is intended to guide the fish towards wattle enclosures surrounded by traps. These are real fish traps that fishermen catch from their boats. The renewal of a Charfya lasts approximately two months.

Figure 36. Charfiya fishing in the Kerkennah Islands, Tunisia. Photographer: Fares Chtioui. (National Institute of Tunisian Heritage INP, 2019)

The Charfya technique was included in 2020 on the representative list of Intangible Cultural Heritage of Humanity by UNSCO. It allows, on the one hand, to highlight island and maritime culture as a transcultural and trans- territorial system of knowledge, know-how and practices, and constitutes, on the other hand, a good example of the harmonious relationship between heritage intangible cultural and its physical environment, with a view to sustainable development (UNESCO, 2020). On the day the Charfya is installed, the sailors' couscous (Fig. 37) is prepared. For this, we use a large container in which we put the chickpeas, beans, raisins, squash, potatoes, octopus.

Figure 37. Sailors' couscous

In a second container, add water and salt in which we cook the pieces of dried meat or Kaddid (AP [2]) with chili powder, coriander and ras hanout (oɪˆ ˆɪɪ) which is a mixture of spices in the form of kebaba (ˆɪ/ɪˆ), cinnamon (Я3 ˆ 3), clove (JjJ ˆɪ&), rosebud Qɪɪ ˆɪˆ), ginger (J?ˆj), fennel (ˆ Ц ˆ O , anise (Sɪˆ <ˆ) and black pepper. The whole thing is cooked a part and will be used to water the steamed couscous which will then be decorated with the vegetables, cooked raisins and chili peppers.

1.5.7. Bezine bel Kaddid or Assida with dried meat and raisins

Soaking the raisins in water before extracting the juice called Makhrouj (ˆɪɪˆ). After grinding and sieving, pieces of dried meat (Kaddid) and lamb ribs are cooked with one or two onions, coriander, caraway, chili pepper, ras hanout and water.

Figure 38. Bezine bel Kaddid

At the end of cooking and adding the makhrouj, we obtain a more or less thick sauce which we pour onto the bezine or assida made from semolina, salt and water which we knead over low heat until obtaining a paste. The Bezine with Kaddid thus prepared is served hot.

1.5.8. Malthouth to Zebib and Karnit

It is a barley couscous sprinkled with an octopus-based sauce to which raisins are added at the end of cooking.

Figure 39. Malthouth at Zebib and Karnit

1.6.Biography of Madame RachidaBen Ezzeddine

Figure 40. Ms. Rachida Ben Ezzeddine

Ms. Rachida Ben Ezzeddine spent three years as a teacher specializing in Kairouan carpet weaving at the Sfax Vocational Training Center. In 1992, it settled in Kerkennah to develop its organic farming whose production, mainly of Asli grapes, is the basis of a diversity of dishes and artisanal pastries. In addition, Ms. Rachida is a member of the Tunisian Women's Union of Ouled Office
On 06/01/1948 , Rachida Ben Ezzeddine was born in machykhet Ouled Bou Ali. She obtained her certificate in tapestry, embroidery and sewing from the Ellahmia school in Sfax in 1964; Holder of a diploma in embroidery, she masters the Kerkennah stitch typical of the scarf or Tarf Kerkennah (Fig. 41), the Nabeul stitch and the Bizerte stitch.

Figure 41. Tarf de Kerkennah brodd by
Ms. Rachida Ben Ezzeddine

Bou Ali in Kerkennah. Very active lady, known through her regular participation in craft fairs as a reference representative of Kerkennese rural women.

Chapter IV

Kerkennah vine varieties

The vine is established throughout the Kerkennah archipelago despite the increasing poverty of the soils and the increasingly severe climatic conditions, associated with the rural exodus of young people. The vines survive in trace amounts in more or less dispersed family orchards and constitute a reservoir of genes of undeniable interest. These vines bear witness to remarkable biodiversity despite continuous transformations and threats dictated by pollution, modernization, socio-economic evolution, human choices as well as climate change. Monitoring and preserving the diversity of these genetic resources in conservation or rehabilitation programs is essential to ensure the maintenance of their adaptive capacities in the face of future environmental changes.

As part of an agriculture concerned with these changes and the environment and in order to preserve the future of our local resources, the contribution of new technologies to the service of agricultural research allows and contributes to the understanding, development and conservation of biodiversity for responsible and sustainable use.

The analysis of the variability of the island's vines was carried out by ampelographic, biochemical, cytological and molecular analyses. The analysis of the morphological characteristics of leaves, shoots, clusters and berries constitutes a powerful ampelographic tool which makes it possible to describe, identify and, once compared, classify vine varieties (Bodor-Pesti *et al.* 2023). Biochemical analyzes of berries provide additional information regarding the characterization of varieties and the determination of their nutritional quality. However, the existence of numerous synonyms and homonyms for cultivars (Harbi Ben Slimane, 2001; Snoussi *et al.* 2004; Calo *et al.* 2008; Cipriani *et al. 2010), means that passport* data are not always sufficient to certify the identities, mainly in terms of distinguishing closely related cultivars, and errors can occur (de Oliveira *et al.* 2020).

As the variability generated by plant growth conditions and health status can also lead to sometimes ambiguous identifications (Cunha *et al.* 2020) and the biochemical characterizations are highly influenced by environmental parameters, the analyzes were then completed. by the use of molecular markers. The latter constitute an effective strategy for more precise identification of cultivars due to the high information content detected directly at the DNA level without environmental influence and from the early stages of plant development (Roychowdhury *et al.* 2014).

Molecular markers are widely used in the genotyping of plant genetic resources. Microsatellite markers SSRs (Simple Sequence Repeats) have gained considerable importance in genetics and plant improvement thanks to their favorable characteristics. One of the major interests of these microsatellites lies in their extremely high polymorphism. This is based on the variation in the number of repetition units, constituting the microsatellite. The analysis of size polymorphism is generally carried out by PCR (Polymerase Chain Reaction) at the level of target microsatellites. The genetic fingerprints (or genetic profiles) generated by this technique make it possible to distinguish varieties within the same species. The major advantage of this molecular technique is to be able to determine the variety of a sample reliably, in any season, at any stage of plant development and from any organ.

Furthermore, molecular approaches allow a massive exchange of data and the use of all available genotyping information. Thus, the characterization of local vine biodiversity is made easier and more accessible.

4.1. Ampelographic description

A so-called ampelographic morphological description was carried out for the range of indigenous vines of Kerkennah. This is a description of different organs of the plant considered at specific stages of vegetative development indicated by the Code for the description of vine varieties.

This code developed in 1983 (OIV, 1983) allows a universal and international "language" for ampelographers. It is used by the International Union for the Protection of New Varieties of Plants (UPOV), by the International Genetic Resources Council (IPGRI) established by the Food and Agriculture Organization of the United Nations (FAO) and by the International Organization of Vine and Wine (OIV).

This numerical code includes 130 ampelographic characters used in the description of vine varieties and clones. Each character is described by a number which provides information on the degree of importance of its expression (Tab. 5).

A given character can be qualitative or quantitative; the qualitative characteristics are coded by numbers, designating the minimum with 1 and no upper limit.

Quantitative characteristics are measurable characteristics according to the following basic scheme:

Table 5: Basic scheme of coding of quantitative ampelographic characters

Character notation	1	2	3	4	5	6	7	8	9
Meaning	Absent or very weak	Very low to low	Weak	Low to medium	AVERAGE	Medium to strong	Strong	Strong to very strong	Very strong

It is important to point out that the measurements relate to a representative sample taken from different parts of the plant and at a well-defined time in its development cycle.

For an easy ampelographic presentation of the varieties studied, we opted for the translation and description coded by text in a sheet relating to each of the varieties studied with an illustration of the essential organs of the variety. The ampelographic information relating to each of the varieties is supplemented by biochemical and molecular data mentioned later in the text (Paragraph 4.4).

Such a presentation allows the reader, experienced ampelographer or simple

Kerkennah enthusiast to proceed to the recognition of the vines of this archipelago.

4.2. Inter-varietal genetic diversity of Kerkennah vines

To be able to effectively and rationally promote and exploit the genetic diversity of the indigenous vines of the Kerkennah Islands, it is essential to have prior knowledge of this diversity, particularly on the genetic level.

Molecular marking makes it possible to reveal varietal diversity and particularly to identify genetically unique populations which are potentially useful for breeders (Aubertin *et al.* 2007; Arroyo-Garria *et al.* 2016). Indeed, genetic resources can constitute reservoirs which contain genes of agronomic interest necessary for their maintenance and sustainability, but also for improvement programs: resistance to diseases or abiotic stress, high yield, better renological quality, taste and nutrition (Flutre *et al.* 2022).

Molecular markers can have several applications. They have thus been widely used for the analysis of the genetic diversity of vines (Aradhya *et al.* 2003; Riahi *et al.* 2010; Ergul *et al.* 2011; De Andres *et al.* 2012; Augusto *et al.* 2021; de Oliveira *et al.* 2022), local intervarietal diversity and analysis of the varietal assortment of germplasm (Dangi *et al.* 2001; Snoussi *et al.* 2004; Ibanez *et al.* 2009; Zinelabidine *et al.* 2010; Laucou *et al.* 2011; de Oliveira *et al.* 2020; Crespan *et al.* 2021), as well as the analysis of phylogenetic relationships (De Andres *et al.* 2012; I§gi, 2019; Cunha *et al.* 2020). The combination of these analyzes with those from historical references made it possible to understand and adjust the history of the varieties/cultivars (Grassi *et al.* 2006;

Crespan *et al.* 2020; Maras *et al.* 2020), and to increase knowledge on the migration of varieties and the flow of genes between populations (Lopes *et al.* 2009; Riaz *et al.* 2018).

Several molecular markers have been developed and applied for the analysis of genetic diversity within the genus *Vitis* , including particularly microsatellite markers (Aradhya *et al.* 2003; Bacilieri *et al.* 2013; Migliaro *et al.* 2022) or SNP markers. (Myles *et al.* 2011; Bacilieri *et al.* 2013; Laucou *et al.* 2018).

In general, genetic diversity results from all the phenomena of DNA modification (mutations, sexual recombination) associated with the effects of natural selection as well as human action. It is based on variations in coding (genes) or non-coding DNA sequences. These variations, whether or not they result in a phenotypic, physiological or biochemical modification, are directly revealed by molecular markers. These markers are therefore neutral indicators of genetic variability to identify interspecific, intervarietal and even intravarietal polymorphism. Microsatellite markers (SSRs) are made up of short tandem repeats of DNA sequence (1 to 6 base pairs) which are dispersed in all Eukaryotic genomes but also in the genome of certain chloroplasts and some mitochondria. These repetitions are found both in coding regions, but more frequently in non-coding regions of the genome.

The overall diversity of cultivated vines has mainly been studied with microsatellite markers (Aradhya *et al.* 2003; Arroyo-Garria *et al.* 2006; Pezzotti *et al.* 2010; Bacilieri *et al.* 2013; Riaz *et al.* 2019), thousands of SNP markers (Myles *et al.* 2011; Bacilieri *et al.* 2013; Laucou *et al.* 2018) or both (Emanuelli *et al.* 2013; Peros *et al.* 2021). Molecular markers (microsatellites), and genetic analysis approaches are used in the identification of vine cultivars (This *et al.* 2004; Daler and Cangi, 2022) for a better understanding of the existing state .

In this work, the 9 nuclear microsatellite marker loci used for the analysis of the diversity of the 8 *Vitis vinifera* L. varieties representative of the wine assortment of the Kerkhenna islands: Asli, Mehdoui, Jerbi, Kohli, Marsaoui, Hamri, Dalia and Tounsi , were developed by previous studies on *Vitis vinifera* and *Vitis riparia* : VVS2 (Thomas and Scott, 1993); VVMD5 and VVMD7 (Bowers *et al.* 1996); ssrVrZAG21, ssrVrZAG47, ssrVrZAG62, ssrVrZAG64, ssrVrZAG79 and ssrVrZAG83 (Sefc *et al.* 1999).

The molecular study of the genetic variability of the island's resources highlighted the allelic assortment of each locus (genetic profile) which showed strong heterozygosity. Heterozygosity for a gene corresponds to the presence of two different alleles of this gene at the same locus for each of the homologous chromosomes. At the DNA level, heterozygosity and allelic richness characterize genetic diversity (Brachet *et al.* 2006).

Overall, the analysis revealed a strong genetic diversity of the varieties analyzed as well as variable allelic richness depending on the accessions. A total of 50 alleles were identified from the 9 microsatellite loci analyzed. Thus, the average number of alleles per microsatellite locus varies between 4 (ZAG83, ZAG47) and 8 (VVMD5, VVS2) with an average of 5 alleles.

Vegetative propagation would also be a key element in the structuring and evolution of the diversity of indigenous vines of the Kerkennah Islands. In fact, it fixes highly heterozygous genetic organizations by considerably limiting the flow of genes between them.

By freezing allelic diversity and connections between alleles at different loci in space and time, asexual reproduction limits the loss of alleles and maintains heterozygosity (Judson and Normark, 1996). It can maintain a greater diversity of alleles but a lower diversity of genotypes compared to sexual populations (Balloux *et al.* 2003).

Based on genetic similarity levels, the 8 varieties were grouped into three groups, showing a total of 7 non-redundant genotypes. With this set of markers, a case of synonymy was highlighted: Mehdoui/Asli varieties. These two accessions, although identified by different morphological characters, and although also very heterozygous, were identical for all the molecular markers analyzed. This morphological variability not translated genetically could be due to mutations during the evolution of these accessions not covered by the loci analyzed.

Furthermore, a previous study involving 3 polymorphic chloroplast microsatellites (cpSSR3, cpSSR5, cpSSR10), indicated intraspecific variability within the varieties analyzed which are grouped into 3 distinct haplotypes (A, C and D) (Snoussi *et al.* 2004). . The first group (haplotype A) brings together the varieties Marsaoui, Tounsi and Jerbi, the second group the varieties Hamri and Dalia (haplotype C), and the third group the varieties Asli, Mehdoui and Kohli (haplotype D).

It should be noted that Tunisian vines belong to the four classes A, B, C and D previously described (Arroyo-Garria *et al.* 2002, Snoussi *et al.* 2004).

Kerkennah vine varieties are grown in areas where wild forms are little present, suggesting that most of them could derive from materials introduced to the region at different historical times. The relatively high inter-varietal morphological polymorphism of vegetatively propagated species, as is the case of the vine, could also be attributed to human selection of new phenotypes followed by clonal multiplication (Ollitrault *et al.* 2003).

The genetic heritage (stock of alleles) analyzed reflects a remarkable capacity for adaptation to environmental conditions typical of the islands.

These conditions, which define a terroir, contribute to giving a unique character, a "typicality" to the vines cultivated and the grapes harvested. Clonal propagation contributes to a specific selection of the most adapted genotypes.

A reference file showing the diversity of vine varieties in this archipelago was produced on the basis of ampelographic, ampelometric, cytological (Harbi Ben Slimane *et al.* 2014) and molecular (Snoussi and Harbi Ben Slimane, 2014) studies. It should be noted, however, that the study of genetic diversity at the intra-cultivar level is more difficult (Calderon *et al.* 2020), and this limitation is based on the expected low variability and on the fact that markers such as SSRs selected from inter-cultivar polymorphisms are less effective in such approaches (Imazio *et al.* 2002; Mercati *et al.* 2016). At the scale of the entire genome, the advent of high-throughput next generation sequencing technologies (NGS) and large-scale genetic association studies (Genome-Wide Association Studies), has made it more efficient SNP identification on a large scale and has considerably optimized access to biodiversity and grapevine improvement (Gambino *et al.* 2017; Vondras *et al.* 2019; Roach *et al.* 2018; Villano *et al.* 2022; Butiuc-Keul and Coste, 2023). High-throughput grapevine genotyping has been successful in distinguishing several clonal genotypes within the French black wine grape variety 'Malbec', proving that the history of propagation has shaped its pattern of genetic diversity (Calderon *et al.* 2020).

That said, the choice of the appropriate genotyping tool must take into account the biological objective, the sample size, the desired resolution and precision as well as the available budget (Villano *et al.* 2022).

Recent and current advances in biotechnology, molecular biology, sequengage technologies, bioinformatics tools and modeling have opened new possibilities for genetic improvement of grapevine.

Besides the importance of molecular marker-based methods and high-throughput NGS-based technologies which have greatly facilitated the development of genotyping techniques, association mapping and genome selection; Advances in these technologies show promising potential in several areas such as genetic transformation or genome editing, disease resistance, resilience to environmental stress and improvement of grape quality and constitute valuable steps towards a production of sustainable, high-quality grapes in view of current and future challenges in viticulture. Such technologies would be a valuable and promising contribution to a more precise and in-depth study of the Kerkennah vines.

4.3. Cytological characterization of Kerkennah vines

The chromosomal enumeration of the 8 representative varieties of indigenous cultivated vines of Kerkennah is carried out using the flow cytometry technique (Harbi Ben Slimane *et al.* 2014).

This technique allows evaluation of the ploidy level of seedlings, by analyzing the quantity of DNA of the nuclear genome by circulating nuclei, previously marked with a dye, in a liquid stream in front of a fluorescent lamp.

The intensity of the fluorescence emitted by these elements one by one is recorded by a photoelectric cell then compared to a control.

The flow cytometer is first calibrated using a diploid control (Fig. 44). The count is transcribed on a histogram in the form of a peak corresponding to a given ploidy rate (Dolezel *et al.* 1989).

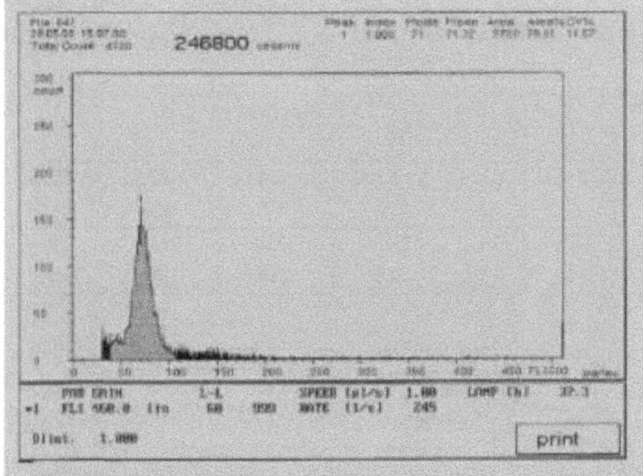

Figure 44. Representation of a diploid model (2n=38)

Flow cryometry analyzes of the varieties Asli, Mehdoui, Razzegui, Jerbi, Kohli, Dalia, Bidh el H'mem showed that all these varieties representative of the indigenous Kerkennah vines are diploid at 2n=38 (Fig. 44).

4.4. Variety sheets of the main vine varieties of Kerkennah

The surveys carried out on the islands made it possible to identify 8 indigenous varieties representative of the varietal assortment of the Kerkennah vines: Asli, Mehdoui, Jerbi, Kohli, Marsaoui, Hamri, Dalia and Tounsi.

ASLI

عسلي ASLI

The variety owes its name to the sweetness of its fruit. Very sweet small grain grape with very few seeds. The grape variety is exclusively grown on the Kerkennah Islands; it is widely used for the manufacture of dried grapes, but it is also used to make a wine very rich in alcohol (16 to 17 degrees) reminiscent of Madeira wine. It ripens towards the end of July. It is a grape variety recommended for making liqueur wines. Very vigorous vine, tree-shaped, with drooping shape.

45

Budding: green, density of hairs layers of the strong end.

Branch: colorful, color of the dorsal surface of the internodes green with red stripes, density of erect hairs at the nodes very low, density of erect hairs at the internodes low.

Tendrils: discontinuous distribution on the branch, very short, average length less than 10 cm. Few rowers.

Adult leaf: small, short size, average length 12 cm, heart-shaped blade with 5 lobes, anthocyanin pigmentation of the main vein of the dorsal side of the blade very weak, blistering of the upper side of the average blade, teeth with convex sides, petiolar sinus slightly open with V-shaped base, upper lateral sinuses with slightly overlapping lobes, density of hairs layered between the veins of the lower surface medium, density of the hairs erect between the veins of the lower surface strong, density of the hairs layered on the veins main veins of the lower side, density of erect hairs on the main veins of the lower side strong.

Flower: hermaphrodite.

Cluster: medium compact, average length 17 cm, average width 10 cm, average weight 215 g. Cylindrical-conical, which can be simple or winged. The cluster has nearly 160 berries

Berry: medium to small size (average length: 15.20 cm, average width: 14.5 cm, non-uniform size, average weight of 1.5 to 2 g, truncated shape, color of the skin green-yellow, golden when ripe, uniform , medium thickness of the skin, barely visible navel, uncolored pulp, quite firm, very sweet.

Berry Biochemical Profile		Sugars (%)			Tartaric acid (g/l)			Maturity Index		
		23.5			1,263			186		
Locus microsatellite	VVMD 5	VVMD 7	VVS 2	VrZAG 21	VrZAG 47	VrZAG 62	VrZAG 64	VrZAG 79	VrZAG 83
Genetic profile (pb)	220-234	232-248	130-142	201-203	159-159	195-199	143-153	249-257	192-195

Plate 1. Asli variety: Adult leaves (upper side, lower side), budding, cluster, flower and seeds (dorsal side, ventral side)

MEHDOUI

MEHDOUI مهدوي

Mehdoui owes its name to Mahdia or Mehdia, a small coastal town in central-eastern Tunisia. The grape variety is found scattered in a few family vineyards. It produces a very beautiful grape with a yellow color, sometimes sincere, with a fairly fine taste, very sought after.

Budding: green, density of hairs layers of the strong end.

Branch: colorful, color of the dorsal surface of the internodes green with red stripes, density of the erect hairs of the nodes and internodes very low.

Tendrils: discontinuous distribution on the branch, short, average length 15 cm.

Adult leaf: medium size, short, average length 12 cm, heart-shaped blade with 5 lobes, anthocyanin pigmentation of the main vein of the dorsal side of the blade very weak, blistering of the upper side of the blade weak, teeth with convex sides, petiolar sinus firm with V-shaped base, firm upper lateral sinuses, with slightly overlapping lobes, density of hairs layered between the veins of the lower surface strong, density of the hairs erect between the veins of the lower surface medium, density of the hairs layered and erect on the main veins of the lower middle surface.

Flower: hermaphrodite.

Cluster: loose, average length 19 cm, average width 10 cm, average weight 30 g. The cluster can be simple or winged.

Berry: small, non-uniform size, average weight 2.3 g, rounded shape, color of the skin green-yellow golden when ripe, uniform, thick skin, umbilicus barely visible, pulp not colored, soft.

Biochemical Profile		Sugars (%)		Tartaric acid (g/l)		Maturity Index		
of the bay		17.9		1.163		154		
Locus microsatellite	VVMD 5	VVMD 7	VVS 2	VrZAG 21	VrZAG 47	VrZAG 62	VrZAG 64	VrZAG 79	VrZAG 83
Genetic profile (pb)	220-234	232-248	130-142	201-203	159-159	195-199	143-153	249-257	192-195

Plate 2. Mehdoui variety: Adult leaves (upper side, lower side), budding, cluster, flower and seeds (dorsal side, ventral side)

JERBI

JERBI جربي

The Jerbi appears to have originated from Vile de Jerba. From there it spread throughout the territory, and particularly in Gabes, Sfax and the Kerkennah Islands. This grape variety is early, found on the market on June 15, productive, giving small grains. It is a white table grape variety which should produce a very good wine because of its very fine taste and its rich sugar content.

Budding: green, glabrous.

Branch: colorful, color of the dorsal side of the internodes green with red stripes, glabrous.

Tendrils: discontinuous distribution on the branch, short, average length 15 cm.

Adult leaf: small size, very short, average length less than 9 cm, pentagonal blade with 5 lobes, teeth with straight sides, open petiolar sinus with U-shaped base, open upper lateral sinuses, density of erect and layered hairs between the veins of the lower surface very weak, main veins of the lower surface glabrous.

Flower: female with reflexed stamens.

Cluster: medium compact, average length 15.5 cm, average width 7.5 cm, average weight 100 g.

Berry: small, uniform size, ovoid shape, color of the skin golden yellow when ripe, uniform, thick skin, barely visible navel, uncolored pulp, firm.

Biochemical Profile	Sugars (%)		Tartaric acid (g/l)		Index de Maturite			
de la baie	26.1		1,050		249			
Locus microsatellite	VVMD 5	VVMD 7	VVS 2	VrZAG 21	VrZAG 47	VrZAG 62	VrZAG 64	VrZAG 79	VrZAG 83
Profile genetique (pb)	222-228	238-246	130-132	190-213	163-163	186-188	137-139	241-249	190-195

Plate 3. Jerbi variety: Adult leaves (upper side, lower side), budding, cluster, flower and seeds (dorsal side, ventral side)

KOHLI

كحلي KOHLI

Kohli or even Kahli exists in most allotment gardens in Kerkennah. Very vigorous strain with great longevity. It produces clusters of great beauty with the advantage of late maturity compared to other varieties of the locality. It is sought after for its crunchy and firm berries which are easy to transport.

Budding: wine red, downy density of layered hairs of the middle end.

Branch: colored, color of the dorsal side of the internodes red, glabrous.

Tendrils: discontinuous on the branch, very long, average length greater than 35 cm.

Adult leaf: small size, average length 13 cm, pentagonal blade with five lobes, teeth with convex sides, petiolar sinus with overlapping lobes, U-shaped base, upper lateral sinuses, density of hairs layered between the veins of the lower surface low, density of the hairs erected between the veins of the lower surface low, density of the hairs layered on the main veins of the lower surface low, density of the hairs erected on the main veins of the lower surface very low.

Flower: hermaphrodite.

Cluster: loose, average length 18 cm, average width 11 cm, average weight 240 g.

Berry: large, uniform size, average weight 4 g, truncated shape, color of the skin initially pink, then becoming dark red, pruine when ripe, dark purplish red, uniform, its very thick skin allows it to to keep for a long time and to export it. Unfortunately its taste is not very fine, it is reminiscent of plum, navel not very visible, pulp not colored, firm, not very sweet.

Berry Biochemical Profile	Sugars (%)	Tartaric acid (g/l)	Maturity Index
	16.3	0.84	197

Locus microsatellite	VVMD 5	VVMD 7	VVS 2	VrZAG 21	VrZAG 47	VrZAG 62	VrZAG 64	VrZAG 79	VrZAG 83
Profile genetique (pb)	228-232	238-250	130-150	190-205	157-163	186-188	139-163	245-245	190-195

Plate 4. Kohli variety: Adult leaves (upper side, lower side), budding, cluster, flower and seeds (dorsal side, ventral side)

MARSAOUI

مرساوي MARSAOUI

> *It is not abundant in the orchards of Kerkhennah. The name of the grape variety probably comes from its town of origin, Marsa, in the northern suburbs of Tunis. Marsaoui is very appreciated for the attractiveness of the cluster of large, firm and crunchy fruits.*

Budding: colored, distribution of anthocyanin pigmentation of the extremity generalized, density of layered hairs of the middle extremity.

Branch: colorful, color of the dorsal surface of the internodes green with red stripes, density of the erect hairs of the nodes and internodes very low.

Tendrils: discontinuous distribution on the branch, average length 20 cm.

Adult leaf: small, short, average length 12 cm, blade pentagonal in shape with 5 lobes, anthocyanin pigmentation of the main vein of the dorsal surface of the blade very weak, upper surface of the blade very weakly bubbled, teeth with convex sides, petiolar sinus very open with U-shaped base, firm upper lateral sinuses, density of hairs layered between the veins of the lower surface very low, density of the hairs erected between the veins of the lower surface medium, density of the hairs layered and erect on the veins main parts of the weak lower face.

Flower: female with reflexed stamens.

Cluster: medium compact, average length 20 cm, average width 15 cm, average weight 650 g.

Berry: medium size, average weight 7 g, non-uniform, rounded, color of the skin green-yellow, golden when ripe, uniform, very thick skin, umbilicus barely visible, pulp not colored, firm.

Profile	Sugars (%)	Tartaric acid (g/l)	Maturity Index
Biochemical of the bay	20.8	1.013	206

Locus microsatellite	VVMD 5	VVMD 7	VVS 2	VrZAG 21	VrZAG 47	VrZAG 62	VrZAG 64	VrZAG 79	VrZAG 83
Profile genetique (pb)	228-236	238-238	144-148	190-205	172-172	186-188	143-143	255-255	190-201

Plate 5. Marsaoui variety: Adult leaves (upper side, lower side), budding, cluster, flower and seeds (dorsal side, ventral side)

HAMRI

حمري **HAMRI**

> *Hamri is fairly well represented in the orchards of Kerkennah; This grape variety has pretty clusters reminiscent of those of Limaoua de Gabes and Bazzoul el Khadem a Raf Raf (Snoussi et al. 2004). Its name would be attributed to the color of its pinkish-red berries. Hamri is renowned for its good production, its resistance to drought and sirocco, its lateness and its capacity to keep on the ground for a long time (Harbi Ben Slimane, 2004; Harbi Ben Slimane and Barka, 2006; Harbi Ben Slimane et al. 2010).*

Budding: bronze coloring, distribution of anthocyanin pigmentation of the tip generalized, density of erect hairs of the tip very low.

Branch: colorful, color of the dorsal side of the red internodes, glabrous nodes and internodes.

Tendrils: discontinuous on the branch, average length 15 cm.

Adult leaf: medium size, average length 15 cm, cuneate blade with five lobes, anthocyanin pigmentation of the main vein of the dorsal surface of the middle blade, blistering of the upper surface of the blade very weak, teeth with rectilinear sides, open petiolar sinus, with U-shaped base, upper lateral sinuses slightly firm, density of erect hairs and layers between the veins of the upper surface of the leaf blade very low, density of erect hairs on the main veins of the lower surface strong.

Flower: hermaphrodite.

Cluster: medium compact, average length 17 cm, average width 8 cm, average weight 260 g.

Berry: large, uniform size, average weight 5.9 g, ovoid shape, color of the skin pink to red, non-uniform, skin of medium thickness, umbilicus barely visible, pulp not colored, firm.

Profile	Sugars (%)	Tartaric acid (g/l)	Maturity Index
Biochemical of the bay	21.8	0.800	279

Locus microsatellite	VVMD 5	VVMD 7	VVS 2	VrZAG 21	VrZAG 47	VrZAG 62	VrZAG 64	VrZAG 79	VrZAG 83
Genetic profile (pb)	232-236	232-244	130-152	190-201	157-172	195-203	139-163	245-255	190-195

Plate 6. Hamri variety: Adult leaves (upper side, lower side), budding, cluster, flower and seeds (dorsal side, ventral side)

DALIA

DALIA داليا

> *The term "dalia" means vine in Arabic. It still symbolizes a flowering branch in Hebrew; it also designates any trellised vine constituting a sort of shelter in the patios of houses.*

Budding: carmine, distribution of anthocyanin pigmentation of the tip generalized, intensity of anthocyanin pigmentation medium, density of lying hairs of the middle tip, density of erect hairs of the tip very low.

Branch: colorful, color of the dorsal and ventral side of the internodes green with red stripes, density of erect hairs of the nodes and internodes very low.

Tendrils: discontinuous distribution on the branch, average length 20 cm.

Adult leaf: small, short size, average length 12 cm, cuneate blade with 7 lobes, anthocyanin pigmentation of the main vein of the dorsal side of the middle blade, blistering of the upper side of the blade very weak, teeth with straight sides, petiolar sinus very open with a U-shaped base, upper lateral sinuses with slightly overlapping lobes, density of hairs layered between the veins of the lower surface medium, density of the hairs erect between the veins of the lower surface strong, density of the hairs layered and erect on the main ribs of the lower side strong.

Flower: hermaphrodite.

Cluster: medium compactness, winged, average length 25 cm, average width 18 cm, average weight 476 g.

Berry: medium size, non-uniform, average weight 3.2 g, ovoid, color of the skin green-yellow, uniform, uncolored pulp, firm, average skin thickness.

Locus microsatellite	VVMD 5	VVMD 7	VVS 2	VrZAG 21	VrZAG 47	VrZAG 62	VrZAG 64	VrZAG 79	VrZAG 83
Genetic profile (pb)	224-236	244-246	130-146	190-201	172-172	199-203	163-163	255-255	192-195

Plate 7. Dalia variety: Adult leaves (upper side, lower side), budding, cluster, flower and seeds (dorsal side, ventral side)

Tounsi

Tounsi تونسي

> *The Tounsi also called Razzegui, M'guargueb or Farrani seems to come from Raf Raf in the north of Tunisia. It was at least in the vineyards of this locality that it was noticed for the first time (Minaingouin, 1905); subsequently, it spread almost everywhere in Tunisia. We are often very impressed by the size of the berries of this grape variety and we cannot fail to think that if we kept this size for all the berries in the bunch, we would obtain an incomparable table grape (Branas, 19~ 4). Tounsi produces a very beautiful grape with firm berries which lend themselves easily to both conservation and transport.*

Budding: reddish, distribution of anthocyanin pigmentation of the extremity generalized, density of layered hairs of the extremity strong.

Branch: strong, colorful, color of the dorsal surface of the internodes green with red stripes, density of the erect hairs of the nodes and internodes very low.

Tendrils: discontinuous distribution on the branch, short, fine, average length 15 cm.

Adult leaf: medium to large size, very short, average length less than 9 cm, pentagonal shaped blade with 3 lobes, anthocyanin pigmentation of the main vein of the dorsal side of the blade very weak, blistering of the upper side of the blade very weak bubbled, teeth with straight sides, open petiolar sinus with U-shaped base, firm upper lateral sinuses, density of hairs layered between the veins of the lower surface low, density of the hairs erected between the veins of the lower surface very low, density layered hairs on the main veins of the lower surface low, density of erect hairs on the main veins of the lower surface very low.

Flower: female with reflexed stamens.

Cluster: medium compact, average length 22 cm, average width 10 cm, average weight 270 g. The cluster is subject to millerandage.

Berry: non-uniform size, average weight 5.5 g, rounded shape slightly depressed at its lower part in the shape of an apple, color of the skin green-yellow golden when ripe, non-uniform, thick skin, visible umbilicus, uncolored pulp, firm and crunchy.

Seeds: large and few.

Profile	Sugars (%)	Tartaric acid (g/l)	Maturity Index
Biochemical of the bay	20.8	1,650	126

Locus microsatellite	VVMD 5	VVMD 7	VVS 2	VrZAG 21	VrZAG 47	VrZAG 62	VrZAG 64	VrZAG 79	VrZAG 83
Genetic profile (pb)	234-236	238-250	130-144	188-203	159-159	186-199	143-163	257-257	192-192

Plate 8. Tounsi variety: Adult leaves (upper side, lower side), budding, cluster, flower and seeds (dorsal side, ventral side)

The Tounsi grape variety is very appreciated by the Tunisian consumer despite its lack of millerandage which greatly depreciates the harmony of the bunch and causes a drop in production. This millerandage results in the heterogeneity of the size of the fruits during their growth phase (Fig. 42), especially when the strain of the grape variety is isolated. This phenomenon is attributed to the infertile pollen of the grape variety (Harbi Ben Slimane *et al.* 2004).

Figure 42. Tounsi millerande

The Tounsi strain is very vigorous (Fig. 43), with a semi-erect shape, very strong, reddish branches, marked grooves and short merithalles. The Tounsi is generally trained in the courtyards of houses. The grape variety is highly appreciated for the size and firmness of its berries. It is the largest of Tunisia's table grapes, which lends itself easily to transport and conservation.

Figure 43. Tounsi palissde strain

4.5. Physiological characterization of Kerkennah vines: Special study of the Asli grape variety

4.5.1. Budburst

Notations of bud break made along Asli pruned wood in irrigated and dry conditions revealed a fluctuation according to their rank of insertion on the branch indicated as follows (Fig. 45):

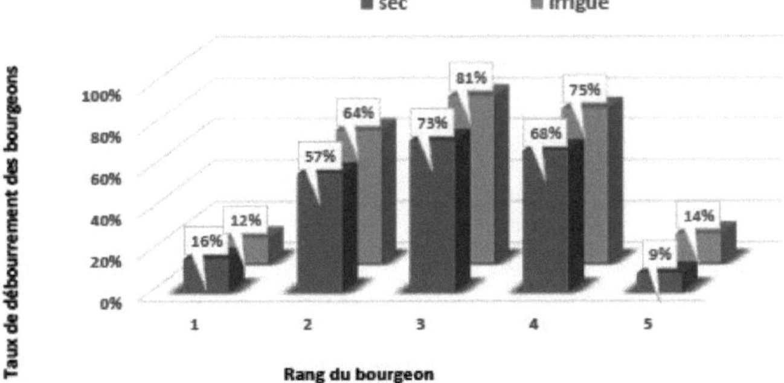

Figure 45. Budburst rate depending on the row on the stick (wood of the year)

The budburst of the first three buds which follow the bud (first bud at the base of the wand, known in the vine by a low rate of expression) is relatively high for the two management methods considered. Beyond the bud of row 5 a remarkable drop in the budburst rate is noted both in irrigated and dry conditions.

4.5.2. Bud fertility

The fertility of the buds along the wood is an important concept for adequate pruning of the grape variety. Fertility is related to the position of the bud on the branch, the internal characteristics of the plant, as well as its vigor which is itself dependent on the pruning applied to two-year-old wood. The notion of fertility is closely related to dry or irrigated management.

The notations recorded in Kerkennah in August 2018 revealed that the fertility of Asli, expressed in average number of inflorescences per branch, was estimated at 0.8 for dry grape varieties and 1.1 for irrigated grape varieties (Fig. 46 A). On the other hand, for Asli, the average number of clusters per plant was 15 under dry conditions and 18 under irrigation (Fig. 46 B).

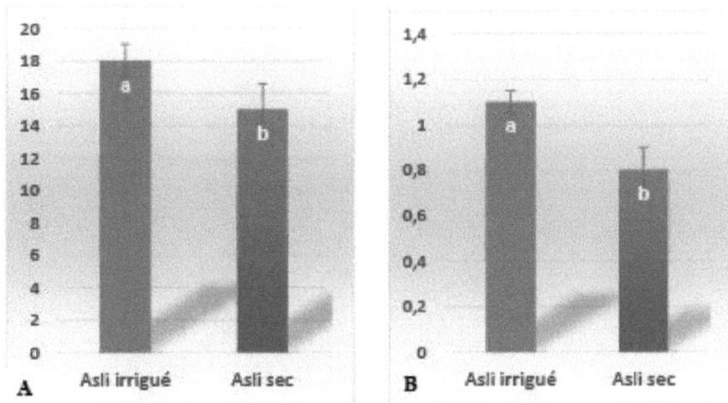

Figure 46. A. Average number of inflorescences per branch **B.** Average number of clusters per plant Leaf area

Measurements carried out in August 2018 revealed that the management method had a very significant effect on the average leaf area of Asli grape varieties grown in the Kerkennah Islands (Fig. 47). Indeed, for plants grown under irrigation, the average surface area of the leaves was significantly greater (350 cm ^{2}) than that of plants grown under rainfed conditions (250 cm ^{2}).

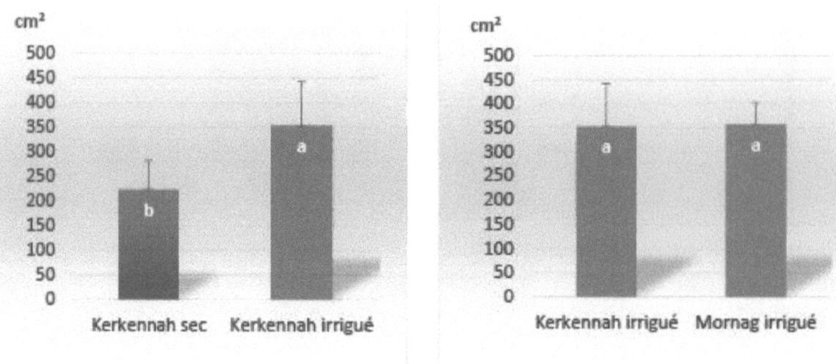

Figure 47: Leaf area of Asli run by rainwater and irrigation in Kerkennah
Figure 48: Leaf area of Asli irrigated in Kerkennah and Mornag

However, comparisons between the average leaf area of irrigated Asli grape varieties in Kerkennah and Mornag (Ben Arous Governorate) did not show significant differences (Fig. 48). Irrigation appears to have a significant effect on the leaf surface of this grape variety regardless of the growing region.

4.5.3. Chlorophyll content of leaves

Chlorophyll is a green pigment that allows photosynthetic plants to carry out photosynthesis.

This process uses sunlight to convert carbon dioxide and water into building blocks for plants. The chlorophyll content of leaves is therefore an indicator of good plant health. This content can be measured by dosing leaves taken from the vine (destructive method) or using various measuring instruments including the SPAD-502 Plus which we used to measure the chlorophyll contents of the Asli grape variety, allowing a quick, easy and non-destructive (Fig. 49).

Figure 49. *In-situ* measurements of chlorophyll content of сёраде Asli a Kerkennah

These measurements have ё ± ё carried out in Kerkennah sur Asli cultivated in irrigim and rainfed as well as in the North in Mornag sur Asli originating from Kerkennah and cultivated in irrigation and this during the month of August 2018 characterized by peaks of temperatures above 35°C, many more frequent in Kerkennah than in Mornag

(Fig. 50).

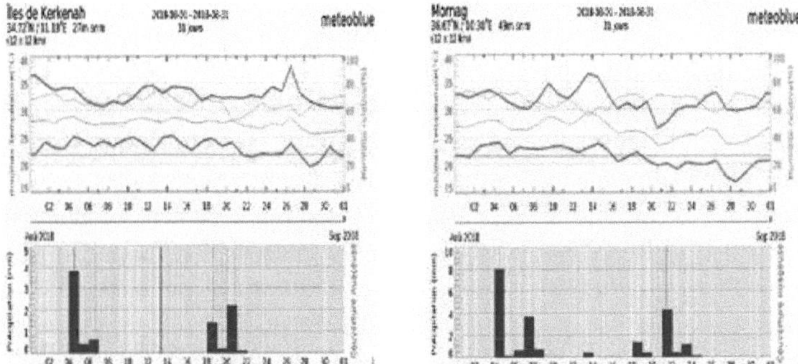

Figure 50. I ктшЩë relative, minimum, maximum and average temperatures and precipitation recorded during the month of August 2018 for the regions of Kerkennah and Mornag (Archives Meteoblue, 2019)

Indeed, the thermal regime of the archipelago is a hot Mediterranean regime, where evaporation is very high, precipitation is low and irregular, leading to massive water shortages amounting to more than 1036 mm/year (Fehri, 2011) .

As a result, water constraints for viticulture in Kerkennah would therefore be a very limiting factor.

After measurements carried out in August 2018, we were able to note that, for the Asli grape varieties grown in the Kerkennah Islands, the method of irrigated or rainfed cultivation had no significant effect on the chlorophyll contents of the leaves (Fig. 51). The absence of irrigation does not seem to affect the chlorophyll content of grape varieties grown under rainfed conditions compared to those grown under irrigation.

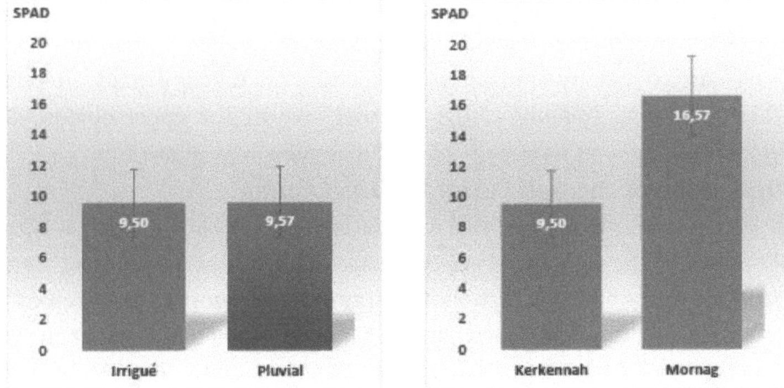

Figure 51: Chlorophyll contents of leaves measured in SPAD units on Asli conducted under rainfed and irrigated conditions in Kerkennah

Figure 52: Chlorophyll contents of leaves measured in SPAD units on irrigated Asli in Kerkennah and Mornag

The water element therefore does not seem to be the most limiting factor for the chlorophyll content. Indeed, morphological adaptations would allow certain species to limit the absorption of light and/or the loss of water (hairiness of the leaves, presence of trichomes, etc.).

Factors common to both driving modes, such as high temperatures, salinity of arable soils and very high global radiation, could explain this degree of equality.

In the Tunisian indigenous varieties Asli and Razegui, Hanana *et al.* (2014) demonstrated that the physiological mechanism of tolerance to abiotic stress is based on their ability to maintain photosynthetic activity despite abiotic stress.

Furthermore, we were able to note that the chlorophyll content of the Asli grape variety cultivated in Kerkennah was significantly lower (-57%) than that of the Asli cultivated in Mornag during the same period and with the same method of irrigation management (irrigated) (Fig. 52). As a result, the photosynthetic potential of Asli would be limited to Kerkennah and would function better in the climatic conditions of Mornag, which are much less restrictive than those of the archipelago. The quality of irrigation water, less salty in Mornag than in Kerkennah, would also play a determining role in limiting the chlorophyll content.

These findings are consistent with the results of Hanana *et al.* (2014), having noted that the Asli grape variety subjected to saline stress (100 mM NaCl) presented a significant reduction in chlorophyll contents (-61%). These authors also emphasize that the adaptation strategies to saline stress of Asli would be less effective than those of the Razzegui grape variety.

Furthermore, structural changes at the leaf level could also explain the significant differences observed between the SPAD values of Asli grown in Kerkennah in the South or in Mornag in the North of Tunisia.

Indeed, a comparative study carried out by Ben Salem-Fnayou *et al.* (2005) between Asli plants grown in the South and North of Tunisia showed structural differences in the leaves. Indeed, the epidermis and the outer wall of the epidermal cells of vine leaves from the South are thicker than those of leaves grown in the North. Observations carried out by transmission electron microscopy made it possible to attribute this thickening (mainly of the external face of the upper epidermis) to the pectocellulose wall, concomitant with an intra-cuticular wax deposition.

4.6. Multiplication of the Asli variety in Kerkennah

This is an initiative to rehabilitate the Asli variety in its original habitat, Kerkennah. Nearly 500 plants of the Asli grape variety were multiplied at the Mornag Agricultural Experimentation Unit under the National Institute of Agronomic Research of Tunisia. For this, cuttings of 60 cm in length and 0.5 cm in diameter were taken from Asli plants in Kerkennah at the end of December 2017. These cuttings (Fig. 53) previously underwent antifungal soaking and conservation of a months in a humid cold room at 4°C in micro-perforated bags.

Soaking the cuttings in exuberone diluted to 50% preceded their gauge stratification until the end of March.

Figure 53. Cuttings at the outlet of the gauge

The stratified cuttings were put into sachets in a substrate consisting of 2/3 peat and 1/3 sand with the addition of 10 g of osmocote per sachet (osmocote being a progressive release fertilizer, capable of to cover the nutritional needs necessary for the plant during the vegetation period).

The cuttings in bags are then placed under shade with control of irrigation and the health state of the plants. Raising the plants continued until the end of January 2018 (Fig. 54).

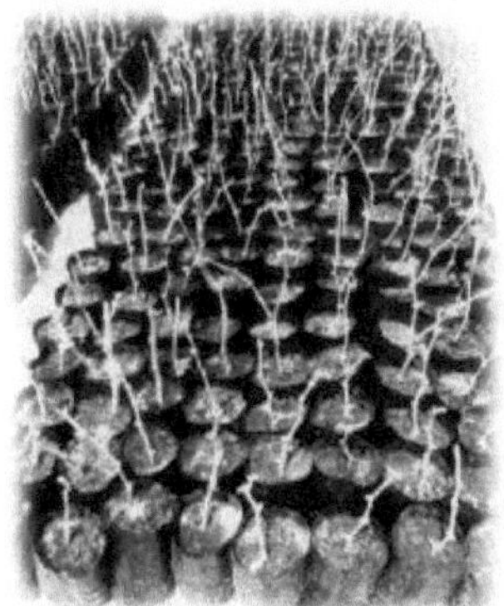

Figure 54. Raising plants under shade

Manual topping was carried out by eliminating the ends of young growing shoots and lateral branches in order to limit the vegetative mass of the plant.

A triple topping was carried out during the vegetative period, thus allowing better control of the vigor of the plant and its health by promoting aeration of the vegetation.

A two-eyed pruning of the plants was done in January 2018 before handing them over to the Organic Agriculture Group located in Ouled Kacem-Kerkennah.

Mr. Nejib Megdiche, president of the aforementioned Group, was responsible for distributing the plants to encourage farmers to replenish the number of Asli plants in Kerkennah, thus allowing the preservation of this grape variety with the sought-after characteristics of the terroir. This action of vegetative propagation of Asli itself constitutes an effective means for the preservation of this grape variety which constitutes the flagship vine of the island.

A program to develop techniques for propagating Kerkennah vines on a larger scale is planned by the Groupement d'Agriculture Biologique-Ouled Kacem-Kerkennah in response to the needs of farmers who are well aware of the importance of such action. The latter hope to rejuvenate their vineyard and ensure the preservation of the Asli grape variety and other ancient Kerkennah grape varieties which are of no less importance.

The Asli root plants (492) were given to farmers in the Zori zone affiliated with the Ouled Kacem Group and to two Asli amateurs in Ras Bounouma and El Ataya (Table 6).

Table 6 : Distribution of Asli root crops to farmers in Kerkennah, 2018

Farmer	Number of plants	Location
Sarsar Ahmed	30	Ezzohri

Megdiche Chaaben	30	Ezzohri
Tarrouch Jamel	30	Ezzohri
Hammeni Daoud	30	Ezzohri
Ezzidi Sami	30	Ezzohri
Hammeni Si'da	30	Ezzohri
Megdiche Habib	30	Ezzohri
El Habib Murad Enneji	30	Ezzohri
El Mechti Sallouha	30	Ezzohri
Megdiche Salah	30	Ezzohri
Gueddiche Mohamed	30	Ezzohri
El Fekki Slim	30	Ezzohri
El Fekki Cherif	20	Ezzohri
El Mechi Abdelmajid	10	Ezzohri
Chahtour Mohamed Khafi	15	Ezzohri
Chahtour Halima	05	Ezzohri
Ezzidi Mohamed	10	Ezzohri
Ezzidi Abdellatif	05	Ezzohri
H'mida Bëlel	10	Ezzohri
Megdiche Megdiche	07	Ezzohri
Echaouech Ennacer	15	Ezzohri
Akrout Tarek	30	Ras Bounouma
Larous Abdelkader	5	El Ataya

Chapter V

Phytosanitary situation of the vines in Kerkennah

The vine, like any crop, is exposed to various pests and diseases described by Galet (1982). Among the main species of hemipteran insects: leafhoppers and aphids, and among the most important viral diseases in the Mediterranean region are the short knot.

5.1.The main pests identified in Kerkennah

5.1.1. Aphids

They are so numerous and diverse that they form a superfamily (*Aphidoideae*) in which polymorphism is one of the key characteristics. The same species can be known in several forms, ranging from winged to able-bodied, sexual or parthenogenetic, oviparous or viviparous. These aphids are globose, oval or spherical in shape and measure between 0.5 mm and 7 mm. The head has a pair of antennae of 3 to 6 segments, compound eyes, a more or less sinuous forehead and a biting-sucking type oral system. The wings have few veins and are most often transparent. The abdomen, of light to dark pigmentation, shiny to matte and oblong to round in shape, can be covered with wax. It is characterized by the presence or absence of a pair of cornicles and a cauda of very variable shape and color depending on the species.

The cornicles, cauda, antennae and pigmentation of the abdomen are the most important identification criteria. The best known species on vines are aphids and phylloxera. Knowing that the indigenous vine varieties of Kerkennah are free growing, the risk of observing phylloxera arises. This is an aphid from the phylloxeridae family which attacks the roots of vines which can cause their death.

The American "Phylloxera of the vine" has gallic forms on leaves and roots giving either virginiparous overwintering radicicolous ones or sexual ones generating winter eggs. In 1863, the damage was so severe that the viability of viticulture in Europe was threatened .

Since then, the vines have been grafted onto American vines resistant to this aphid. The green citrus aphids *Aphis spiraecola* have already been noticed on grape leaves.

In addition, a new species of aphid *Aphis illinoisensis* was introduced from North America and caused serious damage to grape leaves in Sahel vineyards in Tunisia in 2010 (Kamel-Ben Halima and Mdellel, 2010) . We are currently observing this species in several localities of Tunis-ville, Cap Bon and more recently in Kerkennah where we find *A. illinoisensis* on young leaves and vine stems. Both the apters and the wings are brown and shiny forming dense colonies. This species can lead to the formation of abundant sooty mold which degrades the quality of the bunches.

Aphid monitoring: A yellow water trap was installed at the end of April 2016 in Kerkennah and made it possible to identify the aphids circulating in the plot of native vines (Fig. 55).

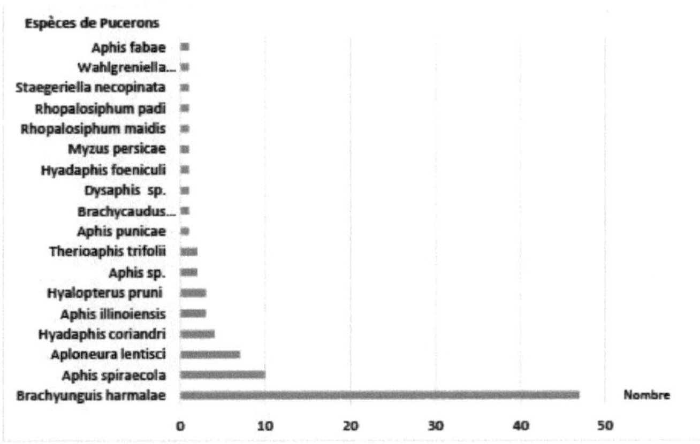

Figure 55. Capture of aphids in the yellow water trap, Kerkennah, 2016

<u>**Species inventory :**</u>

Aphis fabae (Plate 9-16) ,

Aphis illinoiensis (Plate 9-6-7) ,

Aphis punicae (Plate 9-1) ,

Aphis sp., Aphis spiraecola (Planche 9-2) , *Aploneura lentisci* (Planche 9-8) ,
Brachyunguis harmale (Planche 9-14) , *Brachycaudus amygdalinus* (Planche 9-5) ,
Dysaphis sp. (Planche 9-13) , *Hyadaphis coriander* (Planche 9-10) , *Hyadaphis
feniculi* (Planche 9-12) , *Hyalopterus pruni* (Planche 9-4) , *Myzus persicae* (Planche 9-
3) , *Rhopalosiphum maidis* (Planche 9-15) , *Rhopalosiphum padi* (Planche 9-9) ,
Staegeriella necopinata, Therioaphis trifolii (Planche 9-17) , *Wahlgreniella
ossiannilssoni* (Planche 9-11) .

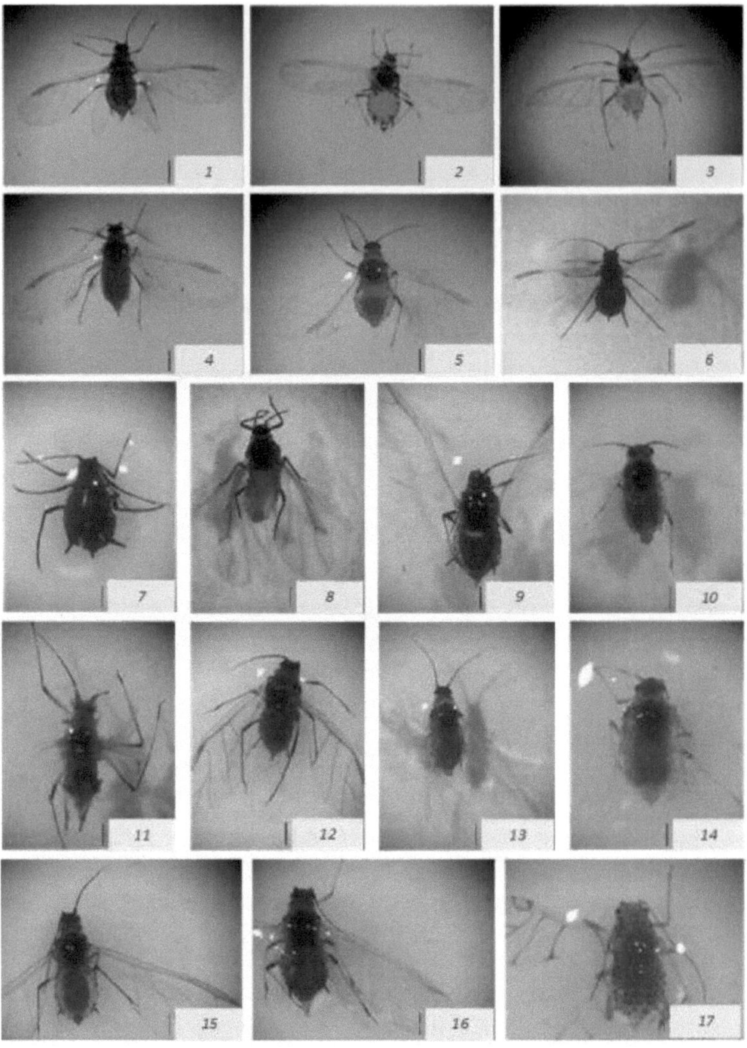

Plate 9. Photos of aphids captured in the Kerkennah yellow water trap, 2016 (1-6; 8-17); 7, wingless *A. illinoisensis* on vine leaves

B. harmalae, A. spiraecola and A. lentisci are the most abundant aphid species in the Kerkennah vineyard with population peaks occurring on May 17, 2016 (Fig. 56).

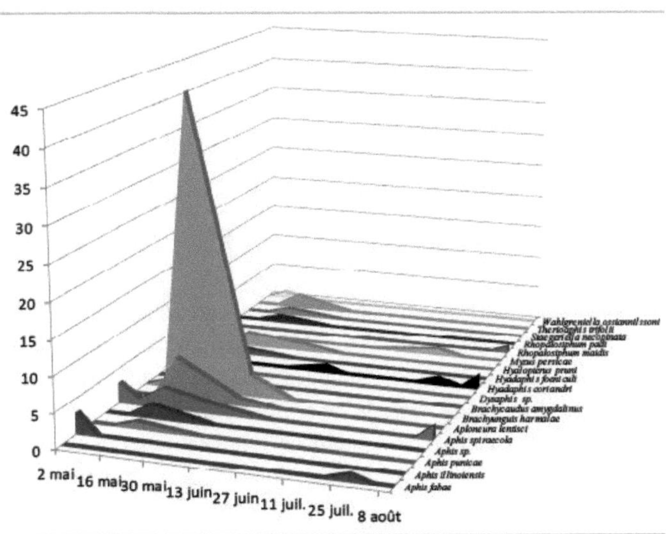

Figure 56. Temporal evolution of aphids in the vineyard, Kerkennah yellow trap, 2016
H. coriandri (condiment aphid), *A. illinoisensis* (invasive vine aphid) and *hyalopterus pruni* (almond aphid) are noted .

At the level of the leaves collected, we were able to identify A. *illinoisensis* (Fig. 57) present at significant densities. Fortunately no aphids of the Phylloxera genus were detected in Kerkennah, a surprising observation given that the native vines are planted in free standing areas.

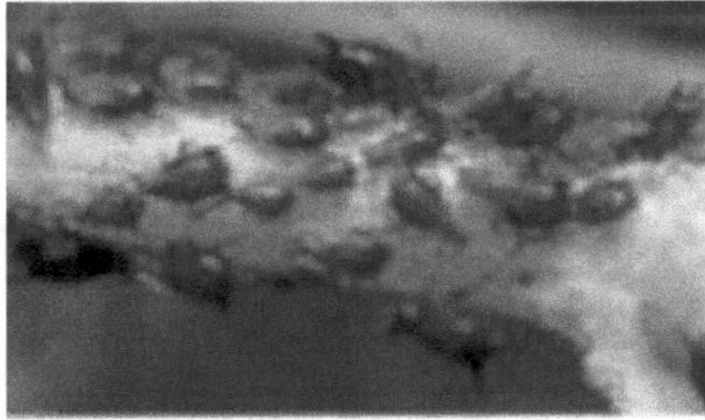

Figure 57. Strong presence of the aphid *A. illinoisensis* on vine leaves

Symptoms and damage : Leaves heavily infested with sap-sucking aphids become yellowish then wilt and curl (Fig. 58). Symptoms may be amplified during the dry season. The honeydew secreted by aphids can lead to the formation of sooty mold, a black fungus that causes grape bunches to deteriorate.

Figure 58. Symptoms of grape leaves infested with sap-sucking aphids

<u>Biological cycle</u> : This complex cycle could be summarized as follows: fertilized winter egg - hatching in the spring of "founder" females giving by parthenogenesis a first generation and also young aphids (virginiparous) by viviparity evolving into apters or wings (several generations in summer) - sexual individuals (sexuparous) in autumn whose females, after mating, produce winter eggs... a cycle during which there will be a change of host or not.

<u>Means of control</u> : Integrated control of aphids may require the introduction of several predators/parasitoids or the use of essential oils to preserve the environment.

5.1.2. Leafhoppers

The leafhopper *Empoasca vitis* has been considered a pest of Tunisian vineyards since the 2000s. *E. vitis* , like aphids, belongs to the order of sap-biting-sucking Hemiptera and to the family Cicadellidae, subfamily Typhlocybinae. They are also small insects, 3 to 4 mm in length, elongated in shape and light green in color (Fig. 59a).

<u>Symptoms and damage</u> : This green leafhopper generates numerous individuals causing the leaves to scorch and dry out through its toxemias, leading to a reduction in sugar yield which is unfavorable for winemaking (Fig. 59b).

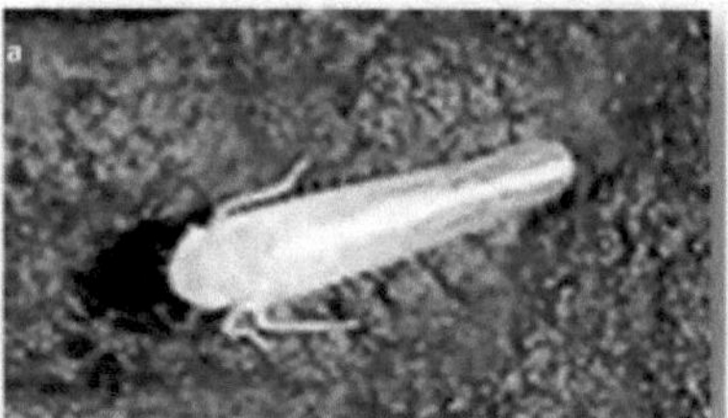

Figure 59. *Empoasca vitis* , adult (a); damage (b)

<u>Biological cycle:</u> *E. vitis* generally produces two to four generations per year. It overwinters as an adult on trees. In spring, she returns to the vineyards. Leafhopper eggs are inserted into the leaves of the plant inside the veins. The larvae are found on the underside of adult leaves where they feed, inducing reddening.

<u>Means of control:</u> The green leafhopper often requires chemical intervention to limit the drying out of the leaves and the reduction in the photosynthetic surface area of the vine. The intervention threshold being 100 larvae on 100 leaves observed during the budburst period.

5.2. Virological status of Kerkennah vines

Among woody species, the vine is considered to be the plant that harbors the greatest number of viruses. The reason is certainly not a particular sensitivity of the plant, but rather its biology which exposes it much more than other species to infections. This relatively long exposure promotes the survival of many viral agents as well as their numerous vectors. Several viruses, often combined, cause serious diseases that can damage the vine and cause significant economic losses.

5.2.1. The short knot

Infectious degeneration is caused by nepoviruses, the most important of which is Grapevine fanleaf virus (GFLV). Court-noue is recognized as one of the most widespread viral diseases in the world and the oldest viral disease of Mediterranean vines. The symptoms of grapevine crown knot appear in spring and differ depending on the strains of GFLV: malformations induced by deforming strains of GFLV and yellowing of leaves induced by chromogenic strains of the virus.

5.2.2. Diagnosis of the virological state of Kerkennah vines

To study the virological state of the Kerkennah vineyard, surveys were carried out at the beginning of May, the optimal period for observing the symptoms of grapevine crown disease. The 4 main wine-growing regions of Kerkennah, namely Ramla, Jouaber, Ouled Ezzedine and Mellita, were visited and 13 plots were explored in different areas of the island.

The vine varieties present in Kerkennah, namely Asli, Mehdoui, Kahli, Tounsi, Marsaoui, Hemri and Jerbi, were the subject of this study.

<u>Visual diagnosis:</u>

A first visual diagnosis was carried out and symptoms of probable viral origin (Plate 10) were observed on branches such as:
- A flattening
- A fasciation translated by a division of the branch into "witch's brooms"
- A shortening of internodes

Plate 10: Symptoms of internode shortening, flattening and fasciation on twigs

These symptoms are observed only in the Mehdoui variety, and are in the majority of cases associated with the same vine. Other symptoms have been observed but which are probably not of viral origin, namely rolling of the leaves, drying of the leaves, chlorosis which could be due to an iron deficiency, etc.

As with most viral diseases, it is generally difficult to establish a diagnosis based on the observation of symptoms.

Indeed, the symptoms caused by the same virus can vary depending on many parameters. In addition, the symptoms expressed can easily be confused with physiological disorders. Therefore, detection techniques must be used to determine the presence of the virus suspected of causing the observed symptoms.

<u>Diagnosis via molecular tools:</u>

Random sampling was carried out, 50 samples were taken from 13 plots in the 4 production zones already mentioned. The sampling was carried out mainly on Asli plants but also on other varieties with or without symptoms of shortening of internodes, fasciations and flattening.

All samples were analyzed with the Reverse Transcription-Polymerase Chain Reaction (RT-PCR) technique for the search for the GFLV virus, using specific primers. Molecular analyzes showed the absence of GFLV in all samples analyzed including those with symptoms of internode shortening, fasciation and flattening. These symptoms, which are generally typical of grapevine crown disease, could have a physiological origin in the Mehdoui variety.

These results were predictable, since a previous study showed that the vine varieties originating from Kerkennah preserved in a collection by INRAT, were found to be free from 6 viruses: viruses 1, 2 and 3 associated with the disease of 1 grapevine leafroll virus (GLRaV-1, -2, -3), grapevine leaf curl virus (GFLV), grapevine mottle virus (GFkV) and mosaic virus 1 'Arabette (ArMV), following DAS-ELISA serological analyzes using polyclonal antisera specific to these viruses (Mahfoudhi *et al.* 2014).

The absence or low rate of infection of Kerkennah vines could be due to the absence of introduction of foreign plant material likely to contaminate indigenous varieties.

Even if the climatic conditions are not optimal, the health state of the vineyard is almost perfect. The epidemic pressure is extremely low. In this regard, we have always insisted that Kerkennians avoid introducing varieties of vines into their localities in order to limit the risks of possible viral contamination.

Capitalization

The vines of Kerkennah constitute a valuable genetic heritage that is an integral part, not only of the agricultural production system, but also of the rich history of the archipelago. The vine is one of the keystones of the cultural and culinary heritage of the Kerkennians.

These vines, and more particularly the Asli de Kerkennah, are endowed with both exceptional quality of fruits and their derivatives but also a capacity for adaptation and resilience to increasingly severe aridity to which adds the absence of viral diseases. All these characteristics constitute advantages which should encourage the maintenance and development of the cultivation of indigenous Kerkennah vines in their original environment using good practices and traditional know-how (Fig. 60).

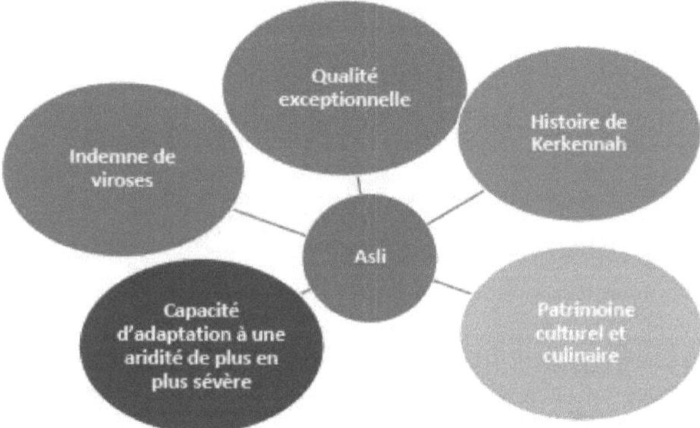

Figure 60. Capitalization of the work Heritage and Wine Traditions in Kerkennah

This work constitutes a contribution to the study of the characterization and diversity of Kerkennah vine varieties. In addition to this fundamental aspect of research, a great deal of effort has been devoted to local practices.

Indeed, this book highlights traditional practices threatened with disappearance and the transmission of traditional know-how relating to the conservation and rational use of biodiversity.

In addition, the richness of the wine-growing heritage of the Kerkennah Islands would have greatly contributed to the sustainability of vine cultivation on the island despite environmental constraints and the smallness of the territory. The preservation of ancient traditional cultivars would make it possible to reduce genetic erosion and produce grapes typical of the islands. Therefore, their maintenance and preservation constitute a priority, mainly because of the predictable and unpredictable hazards induced by nature (climatic or parasitic) and by Man.

The knowledge obtained from our studies on the wine heritage of the Kerkennah Islands is essential for conservation strategies for its biodiversity. Such a rich and unique heritage constitutes an asset for these islands, for their inhabitants but also a very useful support for the selection and improvement programs of the vines of Tunisia.

Nevertheless, collective awareness and joint efforts must be aimed at both the local and national level with a view to massive rehabilitation of the Kerkennah vines. Such rehabilitation will be facilitated by the absence of viral diseases which makes the archipelago a niche to be conserved and protected from any exotic introduction.

One of the most relevant recommendations to defend would be to label the Asli variety of Kerkennah in order to provide significant added value capable of improving the income of Kerkennian farmers. Such labeling could include not only fresh fruit but also raisins, wine, vinegar and all other derivatives and typical dishes.

Such ambitions must first be supported by regional actors such as CTVs (territorial extension units), NGOs (non-governmental organizations), farmers and Kerkennah enthusiasts who must raise awareness and involve decision-makers at the national level.

Finally, support from informed decision-makers would be the basis of an economic, social and cultural revival of Kerkennah. The Ministries to be mobilized would be those responsible for Agriculture, Environment, Research, Culture and Tourism.

Literature cited

HAS

- Allemand-Martin A. 1907. The Kerkennah Islands. Bulletin of the commercial geography society of Paris. 29(1): 83-97.

- Allemand-Martin A. 1940. The Kerkennah Islands: geographical, geological and agricultural overview (continued). In: Monthly Bulletin of the Linnean Society of Lyon, 9 [e] year, n°7-10, pp. 119-125.
DOI: https://doi.org/10.3406/linly.1940.9590.

- Anderson JW, Waters AR 2013. Grape consumption by humans: effects on glycemia and insulinemia and cardiovascular risk factors. Journal of Food Science. 78(1): 11-17. DOI: 10.1111/1750-3841.12071. PMID: 23789931.

- Handles. 2022. Table of nutritional composition of foods Ciqual 2020. National Agency for Food, Environmental and Occupational Health Safety. www.ciqual.anses.fr .

- Aradhya M., Dangl GS., Prins BH., Boursiquot JM., Walker MA., Meredith CP., Simon CJ. 2003. Genetic structure and differentiation in cultivated grape, Vitis vinifera L. Genetics Research. 81(3): 179-192.

- Arroyo-Garria R., Lefort F., de Andres MT, Ibanez J., Borrego J., Jouve N., Cabello F., Martmez-Zapater J.M. 2002. Chloroplasts microsatellite polymorphisms in Vitis species. Genome. 45:1142-1149.

- Arroyo-Garria R., Ruiz-Garcia L., Bolling L., Ocete R., Lopez MA., Arnold C., Ergul A., Soylemezoglu G., Uzun HI., Cabello F., Ibanez J., Aradhya MK., Atanassov A., Atanassov I., Balint S., Cenis JL., Costantini L., Gorislavets S., Grando MS., Klein BY., Mc Govern PE., Merdinoglu D., Pejic I., Pelsy F., Primikirios N., Risovannaya V., Roubelakis-Angelakis KA., Snoussi H. *et al.* 2006. Multiple origins of cultivated grapevine (*Vitis vinifera* L. ssp sativa) based on chloroplast DNA polymorphisms. Molecular Ecology. 15: 3707-3714.

- Arroyo-Garria R., Cantos M., Lara M., Lopez MA., Gallardo A., Ocete CA., Perez
- A., Banati H., Garcia JL., Ocete R. 2016. Characterization of the largest relic Eurasian wild grapevine reservoir in Southern Iberian Peninsula. Spanish Journal of Agricultural Research. 14(3), e0708.

- Augusto D., Ibanez J., Pinto-Sintra A.L., Falco V., Leal F., Martmez-Zapater J.M., Oliveira A.A., Castro I. 2021. Grapevine Diversity and Genetic Relationships in Northeast Portugal Old Vineyards. Plants. 10, 2755.

- Aubertin C., Pinton F., Boisvert V. 2007. Les marches de la biodiversite. IRD Editions, Institut de Recherche pour le Developpement. Paris. 267 p.

B

- Bacilieri R., Lacombe T., Le Cunff L. *et al.* 2013. Genetic structure in cultivated grapevines is linked to geography and human selection. BMC Plant Biology. 13, 25.

- Balloux F., Lehmann L., De Meeus T. 2003. The Population Genetics of Clonal and Partially Clonal Diploids. Genetics Society of America. 164(4): 1635-1644.

- Bassouls C. 2016. Characterization of an indigenous grape variety cultivated in franc de pied : Case of the Asli grape variety from the Kerkennah Islands in Tunisia. End of studies dissertation. Cergy-Pontoise: Higher School of International Agro-Development (ISTOM). France. 114 pp.

- Bell SJ 2011. A review of dietary fiber and health: focus on grapes. Journal of Medicinal Food. 14(9): 877-883.

- Kamel-Ben Halima M., Mdellel L. 2010. First record of the grapevine aphid, *Aphis illinoisensis* Shimer, in Tunisia. OEPP/EPPO Bulletin. 40(2): 191-192.

- Ben Salah M. 2003. The biodiversity of plants cultivated in the Kerkennah Islands. Project report TUN/98/G52/14 (PNUDPMF/FEM-Lions Club Sfax Thyna).

- Ben Salem A., Jemaa R., Guguerli P., Ghorbel A. 2000. Introduction and rehabilitation of vines in the Tunisian Sahara. OIV Bulletin. 835-836: 573,580.

- Ben Salem-Fnayou A., Hanana M., Fathalli N., Souid I., Zemni H., Bessis R., Ghorbel A. 2005. Adaptive anatomical characters of the vine leaf in southern Tunisia. International Journal of Vine and Wine Sciences. 39(1): 11-18.

- Bodor-Pesti P., Taranyi D., Deak T., Nyitraine Sardy DA, Varga Z. 2023. A Review of Ampelometry: Morphometric Characterization of the Grape (*Vitis* spp.) Leaf. Plants. 12, 452. DOI. org/10.3390/plants12030452

- Boulton R. 1980. The relationships between total acidity, titratable acidity and pH in grape tissue. Vitis. 19:113-120.

- Bouzid J., Megdiche J. 2006. Diagnostic report of the southern coastal zone of greater Sfax, extended to the Kerkennah islands, SMAP III Tunisia Project, 144 p.

- Bowers JE, Dangi GS, Vignani R., Meredith CP 1996. Isolation and characterization of new polymorphic simple sequence repeat loci in grape (*Vitis vinifera* L.). Genome. 39:628-633.

- Branas J. 1974. Viticulture. Montpellier, Deh au, Retrieved October 3 2023. 990p. (http://catalog.hathitrust.org/api/volumes/oclc/1534700.html) .

- Brachet S. *et al.* 2006. Reasoned sampling strategies to capture genetic diversity and its structuring in natural populations - Application to conservation management measures. The Proceedings of BRG 6, La Rochelle, France. pp. 211-230.

- Butiuc-Keul A., Coste A. 2023. Biotechnologies and Strategies for Grapevine Improvement. Horticulturae. 9, 62. https:// doi.org/10.3390/horticulturae9010062

C

- Calderon L., Mauri N., Munoz C., Carbonell-Bejerano P., Bree L., Sola C., Gomez- Talquenca S., Royo C., Ibanez J., Martinez-Zapater J.M., Lijavetzky D. 2020. Clonal propagation history shapes the intra-cultivar genetic diversity in 'Malbec' grapevines. bioRxiv. 10.27.356790.

- Calo A., Costacurta A., Maras V., Meneghetti S., Crespan M. 2008. Molecular correlation of Zinfandel (Primitivo) with Austrian, Croatian, and Hungarian cultivars and Kratosija, an additional synonym. The American Journal of Enology and Viticulture. 59: 205-209.

- Cipriani G., Spadotto A., Jurman I., Di Gaspero G., Crespan M., Meneghetti S., Frare E., Vignani R., Cresti M., Morgante M. *et al.* 2010. The SSR-based molecular profile of 1005 grapevine (Vitis vinifera L.) accessions uncovers new synonymy and parentages, and reveals a large admixture amongst varieties of different geographic origin. Theoretical and Applied Genetics. 121(8): 1569-1585.

- CRDA Sfax. 2016. Study on the Kerkennah Islands. Water Resources and Rural Equipment Division, Water Resources District.

- CRDA Sfax. 2000. Annual activity report of the   Regional Commission to the Agricultural Development of Sfax.

- Crespan M., Migliaro D., Larger S., Pindo M., Petrussi C., Stocco M., Rusjan D., Sivilotti P., Velasco R., Maul E. 2020. Unraveling the genetic origin of 'Glera', 'Ribolla Gialla' and other autochthonous grapevine varieties from Friuli Venezia Giulia (northeastern Italy). Scientific Reports. 10:7206 : 1-12.

- Crespan M., Migliaro D., Larger S., Pindo M., Palmisano M., Manni A., Manni E., Polidori E., Sbaffi F., Silvestri Q. *et al.* 2021. Grapevine (*Vitis vinifera* L.) varietal assortment and evolution in the Marche region (central Italy). OENO One. 55(3): 17-37.

- Cunha J., Ibanez J., Teixeira-Santos M., Brazao J., Fevereiro P., Martmez-Zapater JM., Eiras-Dias JE. 2020. Genetic Relationships Among Portuguese Cultivated and Wild Vitis vinifera L. Germplasm. Frontiers in Plant Science. 11: 127.

D

- Daler S., Cangi R. 2022. Characterization of grapevine (V. vinifera L.) varieties grown in Yozgat province (Turkey) by simple sequence repeat (SSR) markers. Turkish Journal of Agriculture and Forestry. 46: 38-48.

- Dangi GS., Mendum ML., Prins BH., Walker MA., Meredith CP., Simon CJ. 2001. Simple sequence repeat analysis of a clonally propagated species: a tool for managing a grape germplasm collection. Genome. 44(3): 432-438.

- De Andres MT., Benito A., Perez-Rivera G., Ocete R., Lopez MA., Gaforio L., Munoz G., Cabello F., Martinez-Zapater JM., Arroyo-Garcia R. 2012. Genetic diversity of wild grapevine populations in Spain and their genetic relationships with cultivated grapevines. Molecular Ecology. 21: 800-816.

- de Oliveira GL., de Souza AP., de Oliveira FA., Zucchi MI., de Souza LM., Moura MF. 2020. Genetic structure and molecular diversity of Brazilian grapevine germplasm: Management and use in breeding programs. PLOS ONE. 15(10): e0240665.

- de Oliveira GL., Niederauer GF., de Oliveira FA., Rodrigues CS., Hernandes JL., de Souza AP., Moura MF. 2022. Genetic Diversity, Population Structure and Parentage Analysis in Brazilian Grapevine Hybrids after Half a Century of Genetic Breeding. bioRxiv. 502144.

- Dolezel J., Binarova P., Lucretti S. 1989. Analysis of nuclear DNA content in plant cells by flow cytometry. Biologia Plantarum. 31: 113-120.

E

- EFSA Panel on Dietetic Products, Nutrition and Allergies. 2004. Opinion of the Scientific Panel on Dietetic products, nutrition and allergies [NDA] related to the

Tolerable Upper Intake Level of Boron (Sodium Borate and Boric Acid). EFSA Journal. 2(8): 80, 1-22. DOI: 10.2903/j.efsa.2004.80.

- Emanuelli F., Lorenzi S., Grzeskowiak L., Catalano V., Stefanini M., Troggio M., Myles S., Martinez-Zapater JM., Zyprian E., Moreira F.M., Grando MS. 2013. Genetic diversity and population structure assessed by SSR and SNP markers in a large germplasm collection of grape. BMC Plant Biology. 13: 1-17.

- Ergul A., Perez-Rivera G., Soylemezoglu G., Kazan K., Arroyo-Garcia R. 2011. Genetic diversity in Anatolian wild grapes (Vitis vinifera subsp. sylvestris) estimated by SSR markers. Plant Genetic Resources. 9(3): 375-383.

- Fehri N. 2011. The palm grove of the Kerkennah Islands (Tunisia), an oasis landscape maritime degradation: natural determinism or anthropogenic responsibility? Physical Geography and Environment. 5:   167-189. DOI: 10.4000/physio-geo. 2011.

- Fehri A. 2014. Kerkhennah, island charm!!! Cercina Center for Research on the Med Islands. Museum and Residence Kerkhennah. Mediterranean Shores Series. No. XIII, 96 p.

- Ferradji A., Goudjal Y., Malek A., 2008. Drying of Sultanine variety grapes using a forced convection solar dryer and a shell-type dryer. Journal of Renewable Energies SMSTS'08 Algiers, 177-185.

- Flutre T., Le Cunff L., Fodor A., Launay A., Romieu C., Berger G. *et al.* 2022. A genome-wide association and prediction study in grapevine deciphers the genetic architecture of multiples traits and identifies genes under many new QTLs. Genes, Genomes, Genetics. 12(7), jkac103.

G

- Galet P. 1982. Les maladies et les parasites de la vigne. Tome 2 : Les Parasites animaux. Montpellier, France, Imprimerie du Paysan du Midi. 1870 p.

- Gambino G., Dal Molin A., Boccacci P., Minio A., Chitarra W., Avanzato CG., Tononi P., Perrone I., Raimondi S., Schneider A. *et al.* 2017. Whole-genome sequencing and SNV genotyping of 'Nebbiolo' (Vitis vinifera L.) clones. Scientific Reports. 7(1): 17294.

- Ghrairi F., Lahouar L., El Arem A., Brahmi F., Ferchichi A., Achour L., Said S. 2013. Physicochemical composition of different varieties of raisins (*Vitis vinifera* L.) from Tunisia. Industrial Crops and Products. 43: 73-77.

- Grassi F., Labra M., Imazio S., Ocete R., Failla O., Scienza A., Sala F. 2006. Phylogeographical structure and conservation genetics of wild grapevine. Conservation Genetics. 7(6): 837-845.

- Hanana M., Hamrouni L., Ben-Hamed K., Ghorbel A., Chedly A. 2014. Comportement et strategies d'adaptation de vignes franches de pied sous stress salin. Journal of New Sciences. 3(4): 29-44.

- Harbi Ben Slimane M. 1999. Study of the genetic variability of indigenous cultivated and spontaneous vines in Tunisia. Doctoral thesis in Biology. Tunis: Faculty of Sciences of Tunis. Tunisia. 160 p.

- Harbi Ben Slimane M. 2001. Ampelography of native cultivated and spontaneous vines of Tunisia. Volume I. INRAT/IPGRI CWANA ISBN 92-9043-502-x.

- Harbi Ben Slimane M. 2004. Ampelography of native cultivated and spontaneous vines of Tunisia. INRAT/IPGRI CWANA. ISBN 92-9043-502-X. 130 p.

- Harbi Ben Slimane M., Chabbouh N., Snoussi H., Bessis R., El Gazzah M. 2004. Study of the germplasm of indigenous vines in Tunisia. Details on the origin of the "Razzegui" millerandage. OIV Bulletin, 77 (881-882): 487-501.

- Harbi Ben Slimane M. 2005. Heritage and wine traditions in Raf-Raf. Ministry of Agriculture and Hydraulic Resources IRESA-INRAT, Official Printing Office, Tunisia. 65 p.

- Harbi Ben Slimane M., Barka M.2006. Presentation of the main characteristics of the "Limaoua" grape variety adapted to the arid Tunisian environment. Meeting International: Resource Management and Biotechnological Applications in Aridoculture and Oasis Crops: Perspectives for valorizing the potential of the Sahara. Jerba, December 25-28, 2006.

- Harbi Ben Slimane M., Trifa Snoussi H., Barka M. 2010. Details on the main characteristics of adaptation of the "Limaoua" grape variety to the Tunisian arid environment. Arid Regions Review. 24 (2).

- Harbi Ben Slimane M., Jedidi E., Bouhlal R., Snoussi H. 2014. Ampelographic, ampelometric and cytological characterization of the diversity of indigenous *Vitis vinifera* L. from Kerkennah in Tunisia. Arid Regions Review. Numero Special - n°35 (3/2014) - Proceedings of the 4th International Meeting "Aridoculture and Oasis Crops: Resource Management and Biotechnological Applications in Aridoculture and Saharan Crops: perspectives for sustainable development of arid zones. 35(3): 237-242.

I

- Ibanez J., Velez MD., de Andres MT., Borrego J. 2009. Molecular markers for establishing distinctness in vegetatively propagated crops: a case study in grapevine. Theoretical and Applied Genetics. 119(7): 1213-1222.

- Imazio S., Labra M., Grassi F., Winfield M., Bardini M., Scienza A. 2002. Molecular tools for clone identification: the case of the grapevine cultivar 'Trammer'. Plant Breed. 121(6): 531-535.

- IPGRI, UPOV, OIV. 1997. Descriptors for Grapevine (Vitis spp.). International Union for the Protection of New Varieties of Plants, Geneva, Switzerland/Office International de la Vigne et du Vin, Paris, France/International Plant Genetic Resources Institute, Rome, Italy. 63 p.

- INP. 2019. La peche a la charfiya aux iles Kerkennah. Institut National du Patrimoine Tunisien Ministere des Affaires Culturelles. Tunisie. https://ich.unesco.org/fr/RL/la-pche-la-charfiya-aux-les-kerkennah-01566.

- I§gi B. 2019. Genetic relationships of some local and introduced grapes (*Vitis vinifera* L.) by microsatellite markers. Biotechnology & Biotechnological Equipment. 33(1): 1303-1310.

J

- Jeszka-Skowron M., Zgola-Grzeskowiak A., Stanisz E., Waskiewicz A. 2017. Potential health benefits and quality of dried fruits: Goji fruits, cranberries and raisins. Food Chemestry. 221: 228-236.
- Judson OP., Normark BB. 1996. Ancient asexual scandals. Trends in Ecology & Evolution. 11(2): 41-46.

K

- Karacali I. 2009. Storage and marketing of horticultural products. Ege University Agriculture Faculty Publication, Bornova, Turkey. Publication no: 494.
- Karadeniz F., Durst R.W., Wrolstad R.E. 2000. Polyphenolic   composition of grapes. Journal of Agricultural and Food Chemistry .   48(11): 5343-5350. DOI: 10.1021/jf0009753. PMID: 11087484.
- Kaur N., Chugh V., Gupta AK 2014. Essential fatty acids as functional components of foods - A review. Journal of Food Science and Technology. 51(10): 2289-2303. DOI: 10.1007/s13197-012-0677-0.

L

- Lachkar A. 2014. Floral biology and productive performance in the local apricot tree (Prunus armeniaca L.). Doctoral thesis in Agricultural Sciences. Sousse: Higher Agronomic Institute of Chott-Mariem, Tunisia.
- Laucou V., Lacombe T., Dechesne F., Siret R., Bruno JP., Dessup M., Dessup T., Ortigosa P., Parra P., Roux C. *et al.* 2011. High throughput analysis of grape genetic diversity as a tool for germplasm collection management. Theoretical and Applied Genetics. 122(6): 1233-1245.
- Laucou V., Launay A., Bacilieri R., Lacombe T., Adam-Blondon AF., Berard A., Chauveau A., de Andres MT., Hausmann L., Ibanez J. *et al.* 2018. Extended diversity analysis of cultivated grapevine Vitis vinifera with 10K genome-wide SNPs. PLoS One. 13(2): e0192540.
- Lerch S., Ferlay A., Shingfield K.J., Martin B., Pomies D., Chilliard Y. 2012. Rapeseed or linseed supplements in grass-based diets: Effects on milk fatty acid composition of Holstein cows over two consecutive lactations. Journal of Dairy Science. 95: 5221-5241.
- Lopes MS., Mendonga D., Rodrigues dos Santos M., Eiras-Dias JE., da Camara Machado A. 2009. New insights on the genetic basis of Portuguese grapevine and on grapevine domestication. Genome. 52(9): 790-800.
- Louis A. 1961. The Kerkennah Islands (Tunisia). Study of Tunisian ethnography and human geography. State Doctorate Thesis, University of Paris, Edit. Institute of Arab Belles-Lettres, Tunis, 26: 418 p.
- Louis A. 1963. The Kerkhennah Islands. Study of Tunisian Ethnography and Human Geography. Publications of the Institut des Belles Lettres Arabes. Tunis. No. 26, 446 p.

M

- Mahfoudhi N., Soltani I.,Digiaro M., Elbeaino T. 2014. Occurrence and widespread distribution of Grapevine virus D in Tunisian grapevines. Journal of Plant Pathology. 96(2): 431-439.

- Maras, V., Tello, J., Gazivoda, A. *et al.* 2020. Population genetic analysis in old Montenegrin vineyards reveals ancient ways currently active to generate diversity in *Vitis vinifera*. Scientific Reports. 10, 15000. https://doi.org/10.1038/s41598-020-71918-7

- Martin A. 2001. Apports nutritionnels conseilles pour la population frangaise. Tec et Doc, 3eme Edition (Retirage 2018). Lavoisier, Paris. 608 p.

- Mercati F., De Lorenzis G., Brancadoro L. *et al.* 2016. High-throughput 18K SNP array to assess genetic variability of the main grapevine cultivars from Sicily. Tree Genetics & Genomes. 12, 59.

- Meteoblue, 2019. https://www.meteoblue.com.

- Migliaro D., De Lorenzis G., Di Lorenzo GS., De Nardi B., Gardiman M., Failla O. *et al.* 2022. Grapevine Non-vinifera Genetic Diversity Assessed by Simple Sequence Repeat Markers as a Starting Point for New Rootstock Breeding Programs. American Journal of Enology and Viticulture. 70: 390-397.

- Minaingouin M. 1901. Note sur la fabrication de raisins secs. Feuille de renseignement. Cite par Louis A. en 1961.

- Minaingouin M. 1905. Etude sur les cepages Tunisiens. Rapport de prospection, Ministere de l'Agriculture de Tunisie, 40 p.

- Murphy D. 2012. Raisin polyphenols and their bioavailability in humans. The FASEB (Federation of American Societies for Experimental Biology) Journal. 26(S1): 421. DOI.org/10.1096/fasebj.26.1_supplement.lb421

- Myles S., Boyko A.R., Owens C.L., Brown P.J., Grassi F., Aradhya M. K., Prins, B., Reynolds, A., Chia, J.-M., Ware, D., Bustamante, C. D., & Buckler, E. S. 2011. Genetic structure and domestication history of the grape. Proceedings of the National Academy of Sciences. 108(9): 3530-3535.   https://doi.org/10.1073/ pnas.1009363108.

N

- Nielsen F.H. 1996. Evidence for the essentiality of Boron. The Journal of Trace Elements in Experimental Medicine. 9: 215-229. https://doi.org/10.1002/(SICI)1520-670X(1996)9:4<215::AID-JTRA7>3.0.C0;2-P

- Naghii M.R., Samman S.   1993. The role of Boron in nutrition and metabolism. Progress in Food and Nutrition Science. 17(4): 331-349. PMID: 8140253.

O

- Office International de la Vigne et du Vin (O.I.V). 1983. Le code des caracteres descriptifs des varietes et especes de Vitis. Ed. Dedon, Paris.

- Ollitrault P., Jacquemond C., Dubois C., Luro F. 2003. Citrus. In: Hamon P., Seguin M., Perrier X. and Glaszmann J-C. (eds.). Genetic diversity of cultivated tropical plants. Montpellier: CIRAD/Science Publishers, Inc., pp. 193-217.

P

Peros J.-P., Cousins P., Launay A., Cubry P., Walker A., Prado E., Peressotti E., Wiedemann-Merdinoglu S., Laucou V., Merdinoglu D., This P., Boursiquot J.-M., & Dloligez A. 2021. Genetic diversity and population structure in Vitis species illustrate phylogeographic patterns in eastern North America. Molecular Ecology. 30(10): 2333-2348. https://doi.org/10.1111/mec.15881

- Pezzotti M., Pe E., Policriti A., & Testolin R. 2010. The SSR-based molecular profile of 1005 grapevine (*Vitis vinifera* L.) accessions uncovers new synonymy and parentages, and reveals a large admixture amongst varieties of different geographic origin. Theoretical and Applied Genetics. 121(8): 1569-1585. https://doi.org/10.1007/s00122-010-1411-9

R

- Rhouma A., Nasr N., Ben Salah M., Allala M. *et al.* 2005. Analysis of the genetic diversity of the date palm in the Kerkennah Islands. Participatory diagnosis of the genetic diversity of the date palm in the Kerkennah archipelago. IPGRI project in Tunisia UNDP/FEM RAB98G31 "Participatory management of date palm genetic resources in the oases of the Maghreb" and Project "conservation of natural resources in the Kerkennah Islands" executed by the Lions Club Sfax Thyna association and financed by PMF/FEM .

- Riahi L., Zoghlami N., El-Heit K. *et al.* 2010. Genetic structure and differentiation among grapevines (*Vitis   vinifera*)   accessions from Maghreb region. Genetic Resources and Crop Evolution. 57:255-272.

- Riaz S., De Lorenzis G., Velasco D., Koehmstedt A., Maghradze D., Bobokashvili Z., Musayev M., Zdunic G., Laucou V., Walker M.A., Failla O., Preece J.E., Aradhya M. and Arroyo-Garcia R. 2018. Genetic diversity analysis of cultivated and wild grapevine (Vitis vinifera L.) accessions around the Mediterranean basin and Central Asia. BMC Plant Biology. 18: 137.

- Riaz S., Pap D., Uretsky J., Laucou V., Boursiquot J.-M., Kocsis L., & Andrew Walker M. 2019. Genetic diversity and parentage analysis of grape rootstocks. Theoretical and Applied Genetics. 132(6): 1847-1860. https://doi.org/10.1007/s00122-019-03320-5

- Roach M.J., Johnson D.L., Bohlmann J., van Vuuren H.J.J., Jones S.J.M., Pretorius I.S., Schmidt S.A., Borneman A.R. 2018. Population sequencing reveals clonal diversity and ancestral inbreeding in the grapevine cultivar Chardonnay. PLoS Genetics. 14(11): e1007807.

- Roychowdhury R., Taoutaou AM., Hakeem KR., Abdel Gawwad MR, Tah J. 2014. Molecular marker assisted technologies for crop improvement. In Crop improvement in the era of climate change, R. Roychowdhury (Ed.), I.K. International Pub. Pvt. Ltd., New Delhi, India. pp. 241-258.

S

- Sefc K.M., Regner F., Turetschek E., Glossl J., Steinkellner H. 1999. Identification of microsatellite sequences in *Vitis riparia* and their applicability for genotyping of different *Vitis* species: Genome. 42: 367-373.

- Schuster M.J., Wang X., Hawkins. T., Painter J.E. 2017. A Comprehensive review of raisins and raisin components and their relationship to human health. Journal of Nutrition and Health. 50(3): 203-216.

- Snoussi H., Ben Slimane Harbi M., Ruiz-Garcia L., Martinez-Zapater JM, Arroyo-Garcia R. 2004. Genetic relationship among cultivated and wild grapevine accessions from Tunisia. Genome. 47(6): 1211-1219.

- Snoussi H., Harbi Ben Slimane M. 2014. Indigenous vines of the Kerkennah Islands between wealth and threatened diversity. Arid Regions Review. Special Issue - n° 35 (3/2014) - Proceedings of the 4th International Meeting "Aridoculture and Oasis Crops: Management of Biotechnological Resources and Applications in Aridoculture and Saharan Crops: perspectives for sustainable development of arid zones. 35(3): 215-220.

- SPA/RAC - UN Environment/PAM, 2019. Management plan for the marine and coastal part of the northern islets of the Kerkennah archipelago - Phase I: diagnostic assessment. By Cabinet Thetis-Conseil, Kheriji A., Limam A., Guellouz S. and Ben Hmida A. Ed. SPA/RAC, Tunis: 79 p.

T

- Thomas M.R., Scott N.S. 1993. Microsatellite repeats in grapevine reveal DNA polymorphisms when analysed as sequence- tagged sites (STSs). Theoretical and Applied Genetics. 86: 985-990.

- Tony J., Alphine J. 2020. Conjugated linoleic acid (CLA): Implications for human health and animal production. The Pharma Innovation Journal. 9(8S): 33-37.

- Trad M., Gaaliche B., Renard C.M.G.C., Mars M. 2012. Quality performance of 'Smyrna' type figs grown under Mediterranean conditions of Tunisia. Journal of Ornemental and Horticultural Plants. 2(3): 139-146.

- This P., Jung A., Boccaci P. *et al.* 2004. Development of a standard set of microsatellite reference alleles for identification for grape cultivars. Theoretical Applied Genetics. 109: 1448-1458.

U

- UNESCO. 2020. Decision of the Intergovernmental Committee: 15. COM 8.b.9. UNESCO, Intangible Cultural Heritage. https://ich.unesco.Org/en/d%C3%A9cisions/15.COM/8.b.9

V

- Van Leeuwen C., Vivin P. 2008. Alimentation hydrique de la vigne et qualite des raisins. Innovations Agronomiques. 2: 159-167.

- Villano C., Cigliano RA., Esposito S., D'Amelia V., Iovene M., Carputo D., Aversano R. 2022. DNA-Based Technologies for Grapevine Biodiversity Exploitation: State of the Art and Future Perspectives. Agronomy. 12, 491.

- Vondras A.M., Minio A., Blanco-Ulate B. *et al.* 2019. The genomic diversification of grapevine clones. BMC Genomics. 20, 972.

W

- Williamson G., Carughi A. 2010. Polyphenol Content and Health Benefits of Grapes. Nutrition Research. 30:511-519. DOI.org/10.1016/j.nutres.2010.07.005.

- World Health Organization (WHO). 1996. Trace elements in human nutrition and health. Geneva, Switzerland. 360 p.

- Zemni H. 2007. Pomological and aromatic characterization of different table and mixed grape varieties of *Vitis vinifera* L. in the pedoclimatic and cultural conditions of their production site. Doctoral thesis in Agricultural Sciences. Tunis: National Agronomic Institute of Tunisia. Tunisia. 163 pp.

- Zinelabidine LH., Haddioui A., Bravo G., Arroyo-Garaa R. *et al.* 2010. Genetic Origins of Cultivated and Wild Grapevines from Morocco. American journal of enology and viticulture. 61(1): 83-90.

yes
I want morebooks!

Buy your books fast and straightforward online - at one of world's fastest growing online book stores! Environmentally sound due to Print-on-Demand technologies.

Buy your books online at
www.morebooks.shop

Kaufen Sie Ihre Bücher schnell und unkompliziert online – auf einer der am schnellsten wachsenden Buchhandelsplattformen weltweit! Dank Print-On-Demand umwelt- und ressourcenschonend produziert.

Bücher schneller online kaufen
www.morebooks.shop

info@omniscriptum.com
www.omniscriptum.com

Printed by Books on Demand GmbH, Norderstedt / Germany